Data Science for Batch Processes

Statistical Learning, Monitoring and Understanding

José M. González-Martínez

José Camacho

Joan Borràs-Ferrís

Alberto Ferrer

WILEY VCH

Authors

José M. González-Martínez
Shell Global Solutions International B.V.
Amsterdam, The Netherlands

José Camacho
University of Granada
and Datharsis
Granada, Spain

Joan Borràs-Ferrís
Universitat Politècnica de València and
Kensight Solutions S.L.
València, Spain

Alberto Ferrer
Universitat Politècnica de València and
Kensight Solutions S.L.
València, Spain

Cover Design: Wiley
Cover Image: © Nordroden/Shutterstock

Typesetting: Lumina Datamatics, Inc

Printing: CPI Group (UK) Ltd, Croydon CR0 4YY

Library of Congress Card No.: applied for

British Library Cataloguing-in-Publication Data A catalogue record for this book is available from the British Library.

Bibliographic information published by the Deutsche Nationalbibliothek The Deutsche Nationalbibliothek lists this publication in the Deutsche Nationalbibliografie; detailed bibliographic data are available on the Internet at <http://dnb.d-nb.de>.

Print ISBN: 9783527326402
ePDF ISBN: 9783527650392
ePub ISBN: 9783527650385
oBook ISBN: 9783527650361

Supplementary instructor material available at www.wiley-vch.de/ISBN<9783527326402>

C9783527326402_290526

Contents

Foreword

Batch processes are ubiquitous across all industrial sectors, including chemical, petrochemical, pharmaceutical, semiconductors, food and beverage, and many others. Their management and optimization are complex, as not only must operational set points be defined (as in continuous processes) but also their evolution across the batch duration. In a way, this flexibility is both a blessing and a curse. There are plenty of degrees of freedom in the batch recipe to enable the production of multiple products and tailor the end properties to specific purposes, but also to introduce unwanted variation, if such degrees of freedom are not properly managed and controlled. Unquestionably, many challenges lie ahead for those who must manage batch processes, which are only properly acknowledged after crossing the bridge from academia to industry.

I was therefore delighted to learn that José M. González-Martínez, José Camacho, Joan Borràs-Ferrís, and Alberto Ferrer have found the time to produce a book devoted to the analysis and monitoring of batch processes – a field to which they have extensively contributed. This is particularly opportune, as there is a clear gap in the academic literature regarding high-quality teaching materials, namely books, for assisting undergraduate and graduate students, as well as researchers and professionals, interested in learning how to analyze data from batch operations.

The book pedagogically gathers the essential components for setting up an analytical workflow for batch data analysis and illustrates how they can be carried out with recourse to examples and MATLAB code. Latent variable models form the data analysis kernel and are therefore carefully introduced, both geometrically and analytically, as well as their endogenous requirements for preprocessing and complexity tuning (selection of the number of latent variables to retain). The more complex topic of synchronization is also covered in detail, both for the offline and online cases. The authors did a remarkable job contextualizing and explaining the different synchronization approaches, illustrating them, and even programming the code to execute them. I believe many of us will be deeply thankful for that! Besides the mainstream topics, the book also features other relevant and practical aspects for data analysis, from which I highlight the attention given to missing data imputation schemes.

I am deeply grateful to the authors for taking the time and for having the relentless resilience to complete the journey of publishing this book. It is now time for the broad community of batch data analysis stakeholders to proceed with their own journey, having such a valuable book as a companion and guide.

Coimbra, August 7 2025 *Marco S. Reis*

Prologue: Challenges for the Third Millennium

Ensuring reproducible and on-specification product quality is one of the main challenges of any industry. This is especially true for batch processes, with some clear advantages compared to continuous processes. Batch processing has feedstock flexibility, a repetitive nature, the ability to handle variations in market demand patterns and to tackle hazardous ingredients or final products. These features make batch processes well suited for the production of low volumes of high value-added products and are in line with the ultimate goals of modern industries: reducing time-to-market and manufacturing costs, complying with regulatory and strict requirements, improving yield and quality, increasing throughput, and reducing energy usage.

The lack of in-depth real-time knowledge about the process state forces manufacturers to operate at conservative, suboptimal, and not intensified regimes. Such practices cause problems such as poor efficiency, lack of process stability, and increased waste of product. To overcome this situation, several challenges must be addressed by the batch process industry, including:

- Better handling of product quality assessment and accomplishment. Quality assessment should not (only) be performed in the final product after batch completion, but also should be integrated (built-in) into the process. This is because in most batch processes, if process anomalies are found too late, little or no correction is possible and the batch product is lost. Therefore, real-time quality control during batch evolution is required. Thus, quality should be achieved by process and product design.
- Better process understanding to facilitate risk-based regulatory decisions and innovation. This is especially important in those industries under strong regulatory norms, e.g. the pharmaceutical industry.

These goals are in line with the guidelines released by the United States Food and Drug Administration (FDA) in the United States in 2004, which encourage industries to detect the presence of variation, calculate its degree, understand its sources, and determine to what extent this may impact the process and ultimately the product features and quality. For this purpose, Process Analytical Technology (PAT) was promoted for designing, analyzing, and controlling production by means of real-time measurements of critical process and materials attributes related to performance and final product quality.[121]

According to the FDA's PAT guidance for industry, the design and optimization of products and production processes within the PAT framework may include the following steps[121]:

1. Identify and measure critical process and raw material attributes related to product quality.
2. Design a process measurement system allowing for real-time (on-, at-, or inline) monitoring of all critical attributes.
3. Develop mathematical models relating the product quality attributes and the measurements of critical attributes in the process and materials.
4. Design process control strategies that provide adjustments ensuring control of all critical attributes.

In a PAT framework, there is continuous quality assurance. The process is continually monitored, evaluated, and adjusted. Multivariate statistical process control (MSPC) is a most valuable tool to be used within this framework. Clearly, the multivariate projection methods embedded into MSPC allow to use real-time measurements and historical databases in a comprehensive way, producing process models relating product quality with critical process variables and materials attributes. These models can be successfully used for process monitoring, fault detection, and control. Moreover, they are very useful for process optimization through data collection and analysis over the historical batches. In this book, we introduce a set of MSPC methodologies to make the most of the historical data collected in a batch process to improve its future performance. The methodologies are nonintrusive, a key advantage to be accepted in an industrial scenario, while revenues may be very important. While not all methodologies will be suitable for a process at hand, we will present strategies for application that will help the implementation by the practitioner in their specific setup. A large amount of practical examples are included throughout the book to facilitate this goal.

This book is supported by a companion website for practitioners. Please refer to the book's page on **Wiley.com** for details on how to access the website. The site provides the following resources, along with instructions for their use:

- Sample datasets in `.mat` format.
- MATLAB Live Scripts that reproduce the examples presented in the book.
- The specific toolbox version required to run and replicate the book's examples.

1

Introduction

1.1 Industrial Batch Processes

Industrial procedures involving raw material and mechanical steps in the manufacture of a product are usually carried out on a very large scale and in different ways. They can be broadly classified as continuous or batch procedures. In both cases, the goal is to operate the process so that desired end-product specifications are met. The achievement of these specifications involves technical decisions at the process operational level, which are related to reaching defined values in a set of relevant process variables through control systems.

Batch processes play an important role in the production of high-added-value specialties in key economic sectors such as manufacturing industry, agriculture, and biotechnology. The manufacture of these products requires a more sophisticated production strategy than the one required in a continuous process. In general, batch processes can be defined as repetitive and finite-duration processes consisting of several actions performed in a sequential manner (see Figure 1.1). A specific mix of raw materials or reactants is initially charged into a vessel. Then, materials are processed for a finite duration with a time-varying control configuration. Optionally, intermediate reactants can be added to the vessel at a proper time. When this is the case, the process may be called a fed-batch process. Once raw materials have been converted into desired products, they are discharged from the vessel and the vessel is cleaned.

This complex type of process has been experiencing a renaissance in the last decades as products-on-demand and first-to-market strategies impel the need for flexible and specialized production methods.[13] Several reasons make batch processes, with their repetitive sequence of charge/process/discharge operations, preferable over continuous processes, where feeding, processing and product extraction are continuous over time. In practice, keeping steady-state conditions over prolonged periods of time is difficult in some processes. For instance, in the biotechnological and pharmaceutical industries that make use of microorganisms, keeping their genetic stability over time is a tough problem. Keeping sterile conditions over prolonged periods of time is another typical problem in many biochemical

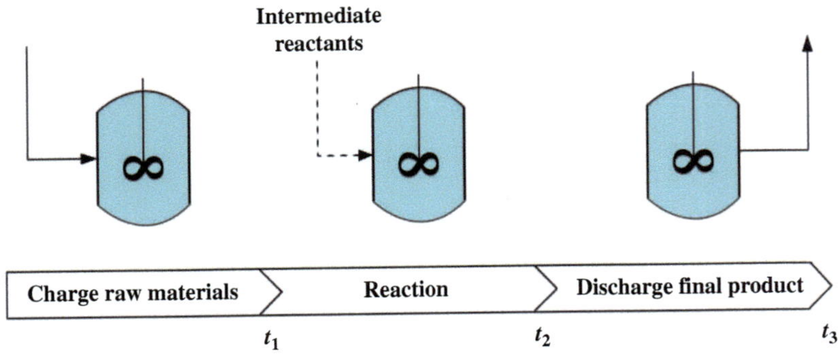

Figure 1.1 Batch process scheme representing the main stages for production.

processes. Also, frequent tuning of process settings for low-volume productions may be economically unfeasible in many processes. As a result, batch processing is usually preferred for low production volumes below 10,000 Mt/year while continuous ones are predominant for production volumes higher than 100,000 Mt/year in industries such as petrochemical, chemical, polymer, and food.[113]

For process control and monitoring purposes, a set of measurements belonging to different process variables is recorded in a batch process. Prior knowledge about the process may determine the sampling frequency to measure the variables, considering for instance the type of variable or the importance of certain phases in the process. Additionally, the quality of a batch can be quantified by collecting online quality measurements provided by probes, as is the case with ammonia probes in wastewater treatment processes. Another possibility to assess the quality of a batch is by analyzing a sample of the final product in a laboratory upon batch completion. These measurements can be seen as end-product quality indices that represent how well the batch was operated. As a consequence of the high cost to obtain these measurements, the frequency at which they are sampled is lower than for engineering measurements (e.g., temperatures, flows, levels). For this reason, and considering that results from the laboratory may be available after some delay, problems raised in a batch may be reproduced in subsequent batches before they are detected and properly diagnosed.

Advances in technology have permitted access to vast databases of engineering data collected in batch processes, which can be leveraged for process monitoring, troubleshooting and optimization. Batches that approximately follow a reference evolution and yield a product within specifications are referred to be under normal operating conditions (NOC). If this NOC can be somehow established, using historical data, as reference trajectories for the process variables, batches which behave abnormally by deviating from these trajectories, so that there is a potential risk of producing out-of-specification product, can be detected long before the quality variables are actually measured. This monitoring capability helps improve profitability in production.

In this book, we introduce a set of methodologies to make the most of the historical data collected in a batch process to improve its future performance. The methodologies are non-intrusive, a key advantage for acceptance in an industrial scenario, while revenues may be substantial. While not all methodologies will be suitable for the process at hand, we will present strategies for application that will help the implementation by the practitioner in his/her specific set-up. Numerous practical examples are included throughout the book to facilitate this goal.

This chapter describes the state-of-the-art in batch process modeling and monitoring. Section 1.2 introduces common types of sensors that can extract information from an industrial process. Section 1.3 briefly describes the major paradigms for modeling batch processes. Finally, in Section 1.4, the modeling cycle of batch processes is reviewed and discussed.

1.2 Types of Sensors

A sensor is any device that reacts to the environment in which it is placed and whose reaction is used to measure the property being sensed. There are many types of sensors to collect data in an industrial process. There are also many versions of each sensor type which may use a different sensing principle or may be designed to operate within different ranges. Some of them can measure physical properties such as temperature, pressure, level, humidity, speed, motion, distance, flow, etc. Others are designed to measure chemicals properties such as pH, oxidation–reduction potential, and concentration or amount of a given chemical species. In bioprocesses, a common type of sensor is a biosensor, which detects specific analytes thanks to a biological component, such as cells, proteins, nucleic acid or biomimetic polymers.

In order to describe and characterize the performance of a sensor, different indices have been proposed[38]:

- **Accuracy** is the degree of correctness when the device is operated under specified conditions. This is typically described in terms of a maximum percentage of expected deviation.
- **Limit of detection** is the smallest measurable input a sensor can signal, e.g., the lowest temperature measured by a thermometer. This concept differs from resolution, which defines the smallest measurable change in input that will produce a small but noticeable change in the output.
- **Linearity** is the degree to which the calibration curve of a device conforms to a straight line.
- **Precision** is the variability among several independent measurements of the same property under the same conditions.
- **Range** is the difference between the minimum and maximum values of sensor output in the intended operating window.
- **Reliability** indicates how well a sensor maintains both precision and accuracy over its expected lifetime.
- **Response time** is the time it takes for the sensor's output to reach its final value. This is a measure of the reaction time to changes in the environment, and must be compared with process dynamics.

- **Selectivity** is the ability of a sensor to measure only one metric or, in the case of a chemical sensor, to measure only a single chemical species.
- **Sensitivity** is the amount of change in a sensor's output in response to a change at the input over the sensor entire range. Provides an indication of a sensor's ability to detect changes. For some sensors, the sensitivity is defined as the input parameter change required to produce a standardized output change.

Some sensors provide a single number for each sample analyzed. In other cases, the provided response is a functional response (profile or spectrum), i.e., a function of some other magnitude such as wavelength (spectrometers) or time (electronic noses and tongues). In these cases, spectra have to be preprocessed (corrected) prior to data analysis in order to remove unwanted systematic variation such as baseline drift, multiplicative scatter effects and profile regions of low information content. Common approaches for preprocessing of spectral data are multiplicative signal correction, standard normal variate correction, Savitzky–Golay smoothing, and first and second order derivatives.[111]

Depending on the way the process sample is analyzed, sensor measurements can be classified as[121]:

- **Offline**, when a sample is analyzed in a laboratory.
- **At-line**, in the production area, during production close to the manufacturing process.
- **Online**, where the measurement system is connected to the process via a diverted sample stream; the sample may be returned to the process stream after measurement.
- **Inline**, where process stream may be disturbed (e.g., probe insertion), and measurement is done in real-time.
- **Noninvasive**, when the sensor is not in contact with the material (e.g., Raman spectroscopy through a window) in the processor, the process stream is not disturbed.

Offline sensors usually provide rich information on chemical properties by using highly specialized analytical instruments such as mass spectrometry, high-performance liquid chromatography or gas chromatography. The main disadvantages of offline sensors are the low-frequency rate due to the analysis cost, the high qualification required to operate the instrument, and the delay in providing the results due to the speed of the instrument and/or the preliminary treatment of the sample.

On the contrary, at-line, online, inline, and noninvasive sensors may provide process information at a fast rate with practically no delays and with simple measurement procedures. Inline sensors may suffer from contamination (e.g., in bioprocesses), leakage through the connections, and build-up of material on them, etc., corrupting the sensor output.

Advancements in biochemistry, chemistry, engineering, and materials science have been important drivers in the development of sensor technologies. Ongoing developments are offering possibilities for new sensors with advanced features, such as greater accuracy, lower cost, and increased reliability.

1.3 Batch Process Modeling

Two main batch modeling paradigms exist in the literature[8]: the mechanistic (or knowledge-based) models inherited from the field of classical mechanics, and data-driven models originally proposed in the field of engineering. The former are developed from the first-principle knowledge of the process in the form of mathematical or biochemical relations, such as reaction kinetics, stoichiometries, and mass/energy balances, that capture the behavior of the process. In contrast, the data-driven models are calibrated solely from historical data. The final goal is to build models that best describe the nature or sources of variation in the process over time. The choice of the model depends on the type of processes and the model purpose: exploratory analysis, optimization, monitoring, control, etc.[72]

The complexity of industrial processes has fostered the development of a large number of model structures and modeling methodologies spanning the whole range between both paradigms, which can be classified into three categories[14]: knowledge-based models, data-driven models, and hybrid models. These models aim to decompose the process data into different sources of variation:

1. A hard part, related to the deterministic behavior or prior process knowledge, such as mathematical relations describing chemical or physical properties.
2. A soft part, formed by the systematic variation that has not been taken into account in the hard part of the model, usually related to unknown sources of variation.
3. A residual part, expected to capture the unsystematic variation of the process, i.e., part of the data variation not explained by the previous parts.

Knowledge-based models decompose process variation into a hard and residual part; data-driven models decompose it into soft and residual parts; and hybrid models include the three parts.

The separation of the different types of variation in process modeling is crucial for process understanding, optimization, monitoring and prediction. Knowledge-based models do not necessarily fit process data and explain batch-to-batch variation as well as data-driven models do. Hence, the use of data-driven models that are capable of capturing not only the known but also the unknown variation sources is highly recommended. However, the use of hybrid models is a good alternative to data-driven models if process knowledge is available. These models use the prior information (hard part) to separate the major variation in data that is mainly generated by known physico-chemical interactions. It permits discovering new phenomena within data that are less abundant (soft part). The remaining variability is kept in the residuals, and is subjected to be modeled separately.

1.3.1 Knowledge-based Models

In this category, both the mathematical models obtained from physicochemical laws – first-principles or fundamental models – and those obtained from the expert knowledge of a process – expert systems – are included. The application of these modeling strategies to batch processes has received a lot of attention by the scientific community for many years, both using dynamical models based on state formulations[51] and rule-based expert systems.[79, 91] These models allow us to estimate the underlying theoretical states of a complex process, such as the growth of bacteria in a certain environment. Take the emergent field of systems biology as an example. In that area, the development of kinetic models of cellular metabolic networks based on quantitative experiments becomes a challenging and primary task for the biotechnological industry. Such models are used to understand the principles that govern cellular behavior and to achieve a predictive understanding of cellular functions for subsequent modeling of fermentations.[69, 156]

Nonetheless, knowledge-based models are seldom suitable for tasks related to monitoring or prediction in batch processes. This is due to the fact that such models are based on assumptions, which may not be accurate for the majority of the process states. Additionally, some unknown reactions may exist. Hence, the uncertainty caused by the unknown knowledge not taken into account in the models may lead to inaccuracies in fault detection and predictions.

1.3.2 Data-driven Models

Current industrial processes, both continuous and batch processes, collect an enormous amount of measurements belonging to hundreds or even thousands of process variables. The analysis of such data in order to distinguish between in-control and out-of-control situations becomes a challenging task. Often, such data are highly correlated (process variables are collinear) with a low signal-to-noise ratio and missing measurements.[95] In this environment, methods to extract the information that explains the majority of the variability of process data are required.[47, 98, 135] Nonetheless, batch data present a number of features which make this extraction cumbersome[166]:

- High dimensionality
- Nonlinear and time-varying dynamics
- Unequalized, uneven, and unsynchronized batch trajectories
- Presence of noise, collinear data, and outliers
- Variables of different magnitude and variance
- Missing data

Empirical multivariate models based on latent variables, such as principal component analysis (PCA) or partial least squares (PLS), coupled with other data processing methodologies developed in the literature, are useful to overcome the above problems, while also providing an analysis tool that is easy to understand from a statistical point of view. Other approaches have been proposed using support vector machines.[122] In batch multivariate statistical process control, a calibration data set composed of NOC batches that are representative of the common process variation is used for model building. Afterward, a test data set is typically used to validate the model parameters for subsequent use in process optimization, monitoring or prediction. For its multiple advantages and general applicability, this is the modeling approach that we will follow throughout the present book.

1.3.3 Hybrid Models

The use of knowledge-based models has the drawback that the unknown part of the process is not represented. Furthermore, some of the underlying assumptions (e.g., reaction kinetics, unknown dynamics, values of the model parameters and objective functions) may not be valid for all the possible states of the process.[89, 110] To overcome this problem, hybrid models combine knowledge-based models, which fit the theoretical behavior, and empirical models, which fit any remaining systematic variation.[48, 142]

Different approaches to decompose the data into the three types of variation (hard part, soft part and residuals) have been proposed in the literature. There are three categories in which these methods can be classified:

- Based on known constraints. There exist general frameworks that enable the imposition of very specific constraints on each type of information, i.e., a priori first principles information, or observed experimental information.[158] This can be done even at different temporal stages of the batch.[157] For batch processes, the basic methodology consists of using the a priori knowledge of a process for building a data-driven model by imposing different constraints, and then modeling the rest of the data using another model, in order to find those unknown sources of systematic variation. Interesting examples can be found in the literature.[72, 127, 170] More recently, grey component analysis (GCA) uses a soft penalty approach to gently force the empirical model toward the direction of the prior information – a chemically or biologically meaningful solution.[178] In summary, these models force the components to be a mixture between the data and the prior information.
- Focused on model parameter estimation. Another option is to use techniques based on introducing the prior knowledge by means of mathematical relations that describe the system behavior or dynamics. The starting point is some specific structure based on first principles, where some functions have to be estimated from data. Different tools can be used to approximate these functions, such as artificial neural networks (NN)[116] or Kalman filters.[11, 151]
- Constraints on the algorithms. Knowledge can also be incorporated in a model by imposing constraints on the modeling algorithms. For instance, some model parameters can be forced to have values within certain ranges, such as in multivariate curve resolution (MCR) or parallel factor (PARAFAC) models.[19, 159]

1.4 Bilinear Modeling Cycle for Batch Process Monitoring

Batch processes often exhibit batch-to-batch variations that are subject to investigation, analysis and monitoring: differences in the chemical composition, types and levels of impurities in raw materials, gradual operational changes, disturbances in the normal processing that affect the quality of the final product, etc. The analysis of data available in each batch is crucial to ensure safe operation, stable product quality and sustainable profit in batch processes. There are four major objectives for analyzing batch data[187]:

i. The analysis of the variable trajectories of historical batches to gain process understanding and troubleshooting past abnormal operating conditions,
ii. The statistical process control of incoming batches either after completion (so-called post-batch process monitoring) or during their progress (real-time process monitoring),
iii. The prediction of the final product quality while the process evolves in real-time, and
iv. The optimization and/or active control of batch operating conditions to reach the desired quality properties of the final product.

The available data in batch processes for statistical analysis mainly fall into three categories[43]:

i. L initial conditions available before the start of the batch for I batches, which are arranged into a two-way array $\mathbf{Z}$ $(I \times L)$. These conditions include raw material properties, preprocessing times such as stage duration and waiting times, information on the shifts, any process measurements taken before the batch starts and expected to have an influence on quality, etc.
ii. J time-varying variable trajectories and/or online analytical sensor responses measured at K different sampling times for all the I batches, which are arranged into the three-way array $\underline{\mathbf{X}}$ $(I \times J \times K)$. Note that under varying sampling rate of variables and/or batches, a preprocessing step of the data will be required to build this three-way array.
iii. M quality and productivity measurements archived after batch completion in the two-way array $\mathbf{Y}$ $(I \times M)$. These quality properties can be collected at K_y different sampling times over the production of each batch, leading to the three-way array $\underline{\mathbf{Y}}$ $(I \times M \times K_y)$ though.

In the design of monitoring schemes, two phases are involved[47]: model building (phase I) and model exploitation (phase II). In the former, understanding the nature of the effects of varying initial conditions and process operating trajectories on the performance of the batches and on the final product quality is pursued.[187] Thereafter, the gained understanding and the statistical models are used to isolate and diagnose past poor operating conditions and set up statistical process control (SPC) schemes for monitoring purposes in the second phase. In model building for process monitoring, a number of steps are typically performed, namely (i) data alignment, (ii) data preprocessing, (iii) transformation of the three-way array to one or several two-way arrays for bilinear batch modeling, and (iv) statistical analysis and monitoring system design. These steps are iteratively repeated, as illustrated in Figure 1.2, to refine the monitoring system of the process prior to model exploitation. The particularities of the processing steps are due to the type of statistical modeling methods used throughout this document, PCA and PLS, which can be considered the state-of-the-art in process modeling. For this reason, we first introduce these methods in Chapter 2 and then move on to detail each of the processing steps.

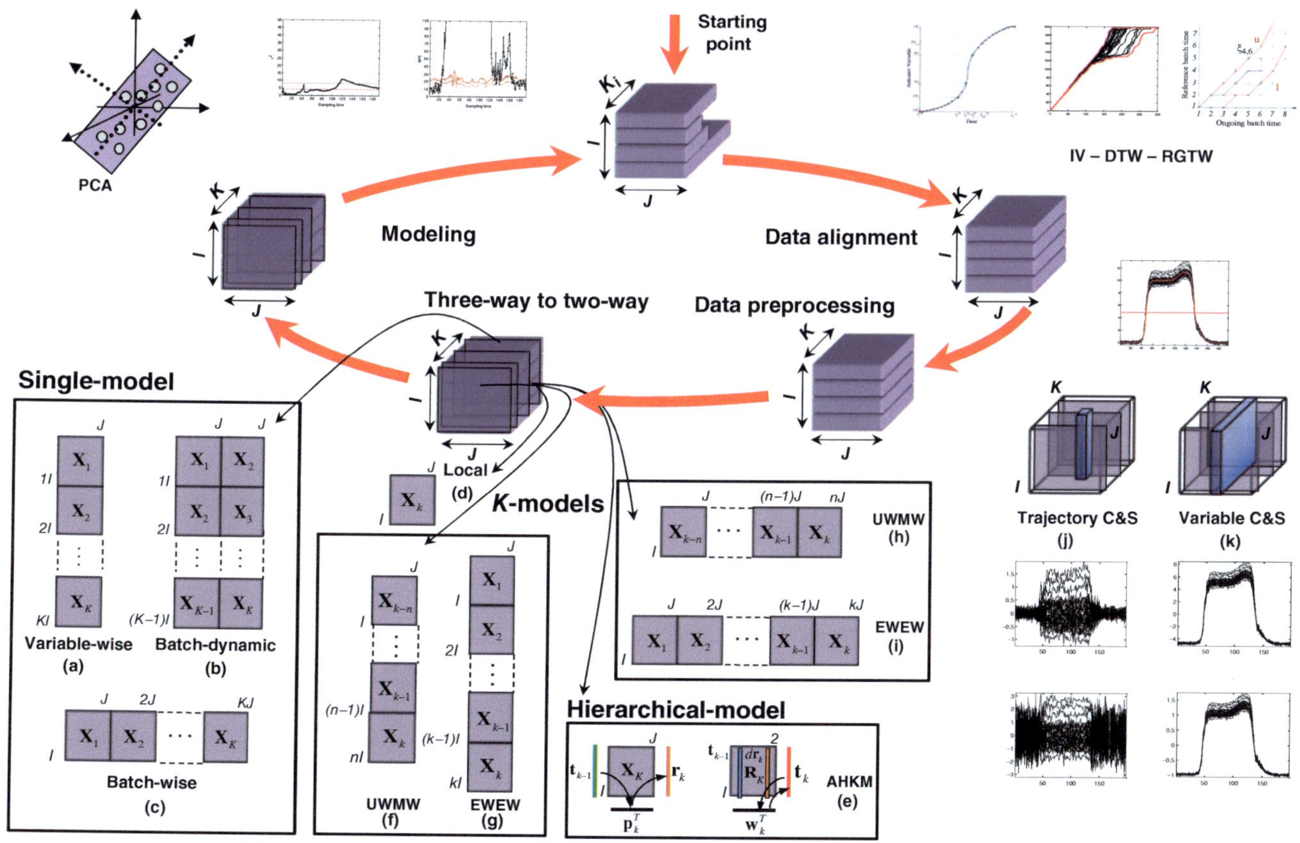

Figure 1.2 Modeling scheme in batch multivariate statistical process control systems based on PCA. *Source*: Reproduced from Ref.[64], with permission of John Wiley & Sons.

In a realistic set-up, the data collected from the process needs to be pretreated to yield data arrays **Z**, $\underline{\mathbf{X}}$, and **Y**. For instance, many of the variables may not be available for all batches and the sampling points for process and/or quality variables may not be the same for different batches or even different variables within the same batch. These mandatory preprocessing steps enable the configuration of data arrays, which is necessary for subsequent modeling and analysis. A fundamental problem is the data alignment of process variables to obtain a three-way data structure $\underline{\mathbf{X}}$ from the data collected through the network of process sensors with multiple sampling rates and for batches of possibly different duration and/or processing pace. This step involves two procedures: (i) batch data equalization and (ii) batch data synchronization. Data equalization involves homogenizing the data within a batch and across batches so that all values reflect process information at the same sampling time and frequency. Batch synchronization, on the other hand, aligns data from different batches to ensure that each sampling time represents similar process information at identical points in the process evolution across batches. The accuracy of empirical multivariate models and subsequent monitoring schemes in terms of fault detection and fault diagnosis is highly dependent on the quality of the equalization and synchronization.[64] This will be discussed in detail in Chapters 3 and 4.

After batch alignment and prior to model calibration, a preprocessing step is needed. Depending on the nature of batch data and the type of model to be fitted, the preprocessing approach may be different.[72] Again, a fundamental choice is the preprocessing applied for three-way matrices. Two main preprocessing methods are widely used in process chemometrics, although they might be referred with different terms: trajectory centering and scaling (trajectory C&S) and variable centering and scaling (variable C&S). The former consists of mean centering and scaling to unit variance the data corresponding to each jth process variable at each kth sampling time. Variable C&S performs mean-centering and scaling to unit variance of the whole data corresponding to each jth process variable. The discussion about which of these two choices is more adequate has been presented in the literature[73, 93, 180] since the two main pioneer research studies in batch monitoring[119, 186] selected one of them each. We address this topic in detail in Chapter 5.

After preprocessing, the transformation of the three-way into two-way data arrays for batch modeling is carried out. This is necessary in order to apply bilinear (two-way) modeling methods, such as PCA or PLS.[1] The different approaches to perform this transformation can be classified into three categories: (i) the single-model approach – e.g., batch-wise (BW), variable-wise (VW) and dynamic model, (ii) the K-models approach – e.g., uniformly weighted moving window (UWMW) and exponentially weighted evolving window (EWEW) models, and (iii) the hierarchical-model approach – e.g., adaptive hierarchical K-models (AHKM). In the single-model approach, the three-way array is unfolded into a single two-way array. The K-models approach is based on generating several bilinear models. The hierarchical-model approach is based on combining the past and current information at each sampling time in a hierarchical bilinear model. Chapter 6 discusses these different methods.

---

1 The statistical monitoring approach based on three-way modeling methods, with very little activity in the research community lately, is not treated in this book.

The final step of the bilinear modeling cycle of batch processes resorts to statistical analysis and the design of the monitoring system. This step is crucial to ensure safe and stable operation, and the production of a high-quality product meeting defined specifications. The statistical analysis and the design of the monitoring system is reviewed in Chapter 7.

2

Data-driven Models Based on Latent Variables

In this chapter, the so-called latent variable models (LVMs) will be introduced. These methods are very useful to analyze large datasets, with a high number of variables, reduced observations, collinear data, missing data, etc. The LVMs of interest in this document are suited for two main purposes: compression and regression.

2.1 Compression

In data compression, we are concerned with the reduction of the volume of data. This may be useful for different purposes: data visualization and understanding, statistical monitoring, missing data recovery, pattern recognition, noise suppression, signal encoding and transmission, and others. In subsequent chapters of this document, we will be interested in the first four applications, from visualization to pattern recognition, which are of paramount importance in the industrial context.

Let the data array $\mathbf{X}_{(\mathbf{M+1})}(I \times V_1 \times \cdots \times V_M)$ contain a set of I experimental conditions, process runs or observations. $\mathbf{X}_{(\mathbf{M+1})}$ is a multidimensional array or tensor with $M + 1$ modes (i.e., ways, indices, or dimensions), the first one representing the observations and the remaining M representing collected variables. Note that each mode m may contain a different number of variables V_m. The data collected for each observation includes a value per combination of variables of the different modes. For instance, each observation of a three-way array $\mathbf{X}_{(3)}$ contains $V_1 \times V_2$ values. An example of a two-way array $\mathbf{X}_{(2)}(I \times V_1)$ is a dataset collected in a sociological study, in which each observation represents an individual and each variable is a feature of this individual, such as age, weight, height, etc. Also, data from a continuous process are typically stored in a two-way array, where each column represents the measurements collected for a process variable, such as temperatures, pressures, etc., and each row represents the data collected for all process variables at a certain sampling time. Data from a batch or fed-batch process are typically three-way: $\mathbf{X}_{(3)}(I \times V_1 \times V_2)$. The data consist of a set of batches. In each batch, a number of process variables are measured at different sampling times during the batch processing.

Thus, the first mode of $\mathbf{X}_{(3)}$ represents the I batches, the second mode represents the V_1 process variables and the third mode represents the V_2 sampling times.

The direct use of data arrays of the general form $\mathbf{X}_{(\mathbf{M+1})}(I \times V_1 \times \cdots \times V_M)$ in many applications is complex due to the size of the data. Because of this, many methods have been defined to reduce the size of the data prior to other operations, such as data modeling and exploratory analysis. A straightforward method to reduce the dimension in the data is variable selection. In variable selection, only a number of relevant variables in each mode are considered. This approach is often too drastic, since some of the omitted data may unexpectedly contain relevant information for the application at hand. Also, traditional variable selection methods neglect the multivariate nature of the data. An alternative, and often more appropriate compression approach, is the use of multivariate compression methods. The general problem of data compression in the context of interest in this book consists in the approximation of the data in $\mathbf{X}_{(M+1)}$ $(I \times V_1 \times \cdots \times V_M)$ using the notion of components or latent variables. Latent variables are linear and orthogonal combinations of the original variables.[1] Provided the latent variables are conveniently identified from the data, the risk of missing important information, as in variable selection is dramatically reduced.

The simplest case of data compression is the two-way case, e.g., to compress the measurements of a continuous process in $\mathbf{X}_{(2)}(I \times V_1)$. Recall that I is the number of sampling times and V_1 the number of process variables. Matrix $\mathbf{X}_{(2)}$ is modeled as follows:

$$\mathbf{X}_{(2)} = \mathbf{T}_A \mathbf{P}_A^T + \mathbf{E}_A \tag{2.1}$$

where $\mathbf{P}_A$ is the $V_1 \times A$ loading matrix, with A the number of latent variables in the model and $A \leq V_1$, $\mathbf{T}_A$ is the $I \times A$ score matrix and $\mathbf{E}_A$ is the $I \times V_1$ residual matrix.

$\mathbf{P}_A$ can be seen as a model of the structure in the data. There are different ways to interpret this model. The columns of $\mathbf{P}_A$ contain the coefficients of the linear combinations of the original variables in $\mathbf{X}_{(2)}$. Each of these columns defines one of the A latent variables in the model. Equation (2.1) can be thought of as a change of variables, so that we create a set of A new variables–the latent variables – from the original V_1 variables, and use the former instead. $\mathbf{T}_A$ contains the values that the I objects take for these new variables.

Equation (2.1) can also be understood in geometrical terms. $\mathbf{P}_A$ spans an A-dimensional subspace contained within the V_1-dimensional space of the original variables. For instance, a one-component model ($A = 1$) from a dataset with two variables (Var 1 and Var 2) can be represented as a one-dimensional line contained in a two-dimensional plane. Each of the I objects in $\mathbf{X}_{(2)}$ can be represented as a point in the V_1-dimensional original space, and its corresponding score as a point in the A-dimensional subspace. $\mathbf{P}_A$ conforms an orthonormal basis of the

1 For the sake of simplicity, we will restrict ourselves to the case, where latent variables are orthogonal and linear combinations of the original variables. Non-orthogonal[81] and nonlinear[41] LVMs are beyond the scope of this book.

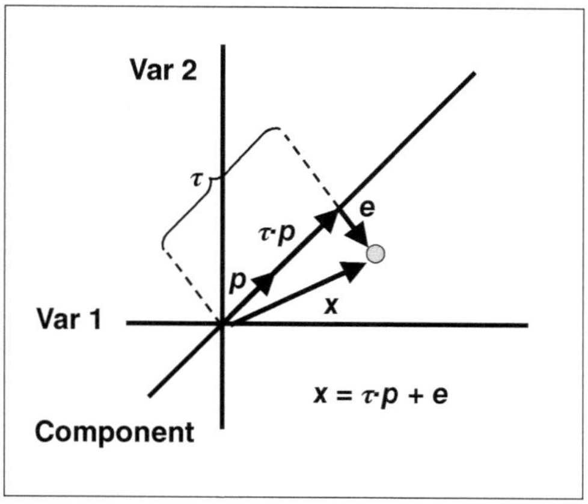

Figure 2.1 Geometrical illustration of the component notion.

A-dimensional subspace[2] and scores are obtained by projecting the observations onto that basis.

The geometrical interpretation of Eq. (2.1) is illustrated in Figure 2.1 for one single observation ($\mathbf{x}$) and again the simplest case of two original variables and a one-component model. The observation $\mathbf{x}$ is represented as a gray point, or as the vector that connects the origin of coordinates to that point. The single loading vector $\mathbf{p}(2 \times 1)$ conforms the basis of the 1-dimensional latent subspace. The score τ is the magnitude of the resulting point in the $\mathbf{p}$ direction when $\mathbf{x}$ is projected onto the model subspace. Just like $\mathbf{x}$, the residual error $\mathbf{e}$ can be seen as a vector in the original space. The quadratic norm of $\mathbf{e}$ is the square of the Euclidean distance of the observation $\mathbf{x}$ to the model.

Factors $\mathbf{T}_A$ and $\mathbf{P}_A$ in Eq. (2.1) are obtained from the dataset $\mathbf{X}_{(2)}$ by using an algorithm that optimizes a certain criterion or objective function. We call "model training" or "model calibration" this process of computing the scores and loadings, and we refer to $\mathbf{X}_{(2)}$ as the "calibration data." The objective function used throughout this book is to minimize the sum of squares of the elements in $\mathbf{E}_A$. This is called PCA,[85] as discussed in the next section. The projection -scores- of a new observation in the components subspace is computed following:

$$\boldsymbol{\tau}_{\text{new}}^T = \mathbf{x}_{\text{new}}^T \mathbf{P}_A \tag{2.2}$$

where $\mathbf{x}_{\text{new}}^T$ is a $1 \times V_1$ vector (one-way array) representing the new observation and $\boldsymbol{\tau}_{\text{new}}^T$ is a $1 \times A$ vector representing its projection into the latent subspace.

This idea of components or latent variables can be generalized to the multi-way case $\mathbf{X}_{(\mathbf{M}+1)}$ following one out of two possible approaches: the unfolding[185] and the

2 The length – the quadratic norm – of the columns of $\mathbf{P}_A$ is not constrained in Eq. (2.1). Thus, if one column of $\mathbf{P}_A$ is multiplied by a quantity k and the corresponding column of $\mathbf{T}_A$ is divided by the same quantity the product $\mathbf{T}_A\mathbf{P}_A^T$ remains the same. A typical convention to correct for this non-uniqueness is to scale the columns of $\mathbf{P}_A$, that is $\mathbf{p}_a$ for $a = \{1, \ldots, A\}$, to unit Euclidean norm ($|\mathbf{p}_a| = \sqrt{\sum_{v_1=1}^{V_1} p_{a,v_1}^2} = 1$). Following this convention, $\mathbf{P}_A$ becomes an orthonormal basis of the latent subspace.

multi-way analysis.[150] The unfolding method is based on transforming $\mathbf{X}_{(M+1)}$ into a two-way array $\mathbf{X}_{(2)}$ and then applying the decomposition in Eq. (2.1). The typical and most studied transformation is to maintain the mode of the observations and combine the rest of the modes into a single one. Thus, the unfolded mode contains one unfolded variable per each possible combination of variables of the different modes. Performing such a transformation, the dimension of $\mathbf{X}_{(2)}$ becomes $(I \times (V_1 + V_2 + \cdots + V_M)$, with one row of data per observation. An example of this transformation for a three-way array of data is shown in Figure 2.2.

The multi-way or N-way analysis consists of developing one set of components for each of the M modes of variables in $\mathbf{X}_{(M+1)}$. Thus, a loading matrix per mode is obtained, possibly with different number of components. Also, a core matrix to model interactions among modes is considered. From the M loadings matrices and the core matrix, the scores are obtained. Examples of multi-way models are parallel factor analysis (PARAFAC)[33, 74] and Tucker-3.[164]

This book is concerned with the application of LVMs in the analysis of batch processes. Data from a batch process are a three-way tensor. Thus, unfolding or three-way analysis methods need to be used. In this book, only the unfolding approach will be considered. There are good reasons for this. Two-way analysis methods are simpler and much faster to calibrate than their three-way counterparts. Thus, the former can be efficiently used for computationally demanding pattern recognition problems, as it will be shown. Furthermore, comparisons in the literature of the two approaches did not show a clear benefit in the use of three-way analysis[104, 149, 180, 182] instead of the unfolding approach. For more detailed information regarding three-way analysis methods, refer to Ref.[150] In Chapter 6, a detailed description of the most used unfolding methods and their implications in batch data modeling can be found.

For the sake of easy understanding, from here onward, letters j and J will be used for the second mode of the two-way data array, and letters k and K for the third mode. Also, $\mathbf{X}$ will stand for a two-way array $\mathbf{X}_{(2)}$ and $\underline{\mathbf{X}}$ for a three-way array $\mathbf{X}_{(3)}$.

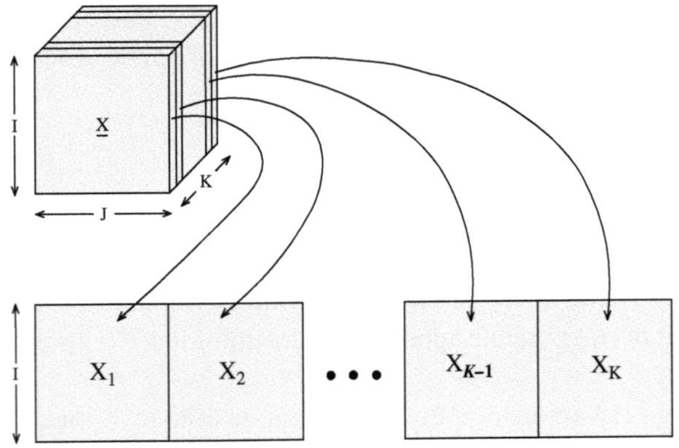

Figure 2.2 Unfolding from three-way to two-way data by maintaining the first mode. $\underline{\mathbf{X}}$ stands for the three-way array $\mathbf{X}_{(3)}$.

Example 2.1: The pipe network example

This example can be reproduced with the MATLAB live script called "ex2_1.mlx." In Figure 2.3, the data simulation approach, taken from,[32] is shown. Three liquids are mixed in two pipes. Data are simulated according to the following equalities:

$$F12 = (4/5)F1 + (1/5)F2$$
$$F123 = (16/25)F1 + (4/25)F2 + (5/25)F3$$

The flows F1, F2, and F3 are generated with independent and normally distributed random values with a 0 mean and unit standard deviation. This represents that the average of the flows has been previously removed from the data measured from the pipelines, and the remaining data represent the variability around the average value. Samples generated are not autocorrelated. This would be the case for data collected separately enough in time so that dynamics can be neglected. The main input flow as well as the flows in the intermediate pipes are measured: $\mathbf{X} = \{F1, F12, F123\}$. One hundred samples are collected from this simulation, yielding a 100×3 two-way array of data.

Let use consider the following two-way data array $\mathbf{P}_2$ represents a model of the data:

$$\mathbf{P}_2 = \begin{bmatrix} \frac{1}{\sqrt{2}} & \frac{1}{\sqrt{2}} \\ \frac{1}{\sqrt{2}} & -\frac{1}{\sqrt{2}} \\ 0 & 0 \end{bmatrix}$$

The scores are computed according to $\mathbf{T}_2 = \mathbf{X}\mathbf{P}_2$. Subsequently, an estimate of $\mathbf{X}$, noted $\widehat{\mathbf{X}}_2$, can be obtained as $\widehat{\mathbf{X}}_2 = \mathbf{T}_2\mathbf{P}_2^T$. Matrix $\mathbf{P}_2$ transforms the original three variables in a pair of orthogonal components (also known as latent variables). These

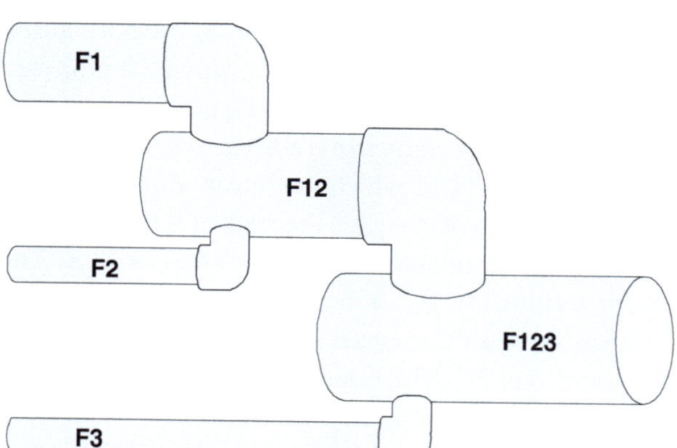

Figure 2.3 Example for data generation.[32] Three liquids are mixed in two pipes. The process variables are the input flows $F1$, $F2$, and $F3$ and the flows of the intermediate pipes $F12$ and $F123$. For illustrative purposes, let us assume no delays in the process, so that the output flow depends solely on the current input flows. *Source*: Reproduced from Ref.[32], with permission of Elsevier.

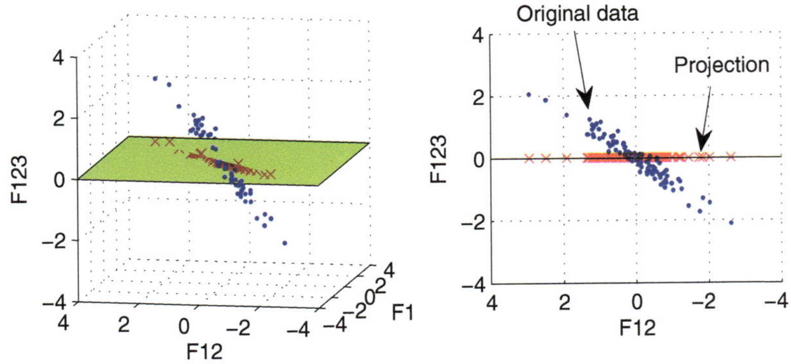

Figure 2.4 Display of the results of Example 2.1. Observations of $\mathbf{X}$ are shown as points. Observations of $\widehat{\mathbf{X}}$ are shown as crosses.

components span a plane in $F123 = 0$. The result of the projection can be seen in Figure 2.4. The residuals $\mathbf{E}_2 = \mathbf{X} - \widehat{\mathbf{X}}_2$ are the difference between each point and its projection to the plane.

2.2 Principal Component Analysis

Principal component analysis (PCA) is suited for the analysis of two-way matrices of the form $\mathbf{X}(I \times J)$, composed of I observations on J variables. It is generally accepted that PCA has its origins in the work by the British statistician Karl Pearson at the beginning of the twentieth century, "On Lines and Planes of Closest Fit to Systems of Points in the Space".[124] Thus, returning to the discussion in the previous section, PCA has its origin in its more intuitive geometric interpretation. The term "Closest Fit" stands for the solution which minimizes the sum-of-squares of residuals. Therefore, PCA follows Eq. (2.1) with $\mathbf{P}_A$ set so that the sum-of-squares of $\mathbf{E}_A$ is minimum. This is equivalent to state that the cost function in PCA is to maximize the sum-of-squares in the product $\mathbf{T}_A\mathbf{P}_A^T$. Very often, the loading vectors in $\mathbf{P}_A$ are normalized to unit length. Thus, PCA finds the A-dimensional subspace in which the projected observations, that is $\mathbf{T}_A$, present maximum sum-of-squares (Figure 2.5). For that, PCA extracts the orthogonal directions of maximum sum-of-squares in the variables space, the so-called principal components (PCs), which are linear combinations of the variables. Note that the maximum value of A equals the rank of $\mathbf{X}$.

The direction of the PCs is obtained from the eigenvectors of $\mathbf{C}_X = \mathbf{X}^T\mathbf{X}$. The eigenvectors are arranged in the columns of $\mathbf{P}_A$, ordered by their corresponding eigenvalues. The eigenvalue associated to a PC measures the sum-of-squares of the projections of the observations onto this PC. The nonlinear iterative partial least squares (NIPALS) algorithm[183] computes the PCs in order, so that the discarded directions do not have to be computed. Also, the singular value decomposition or the eigen-decomposition (among others) can be used to derive the PCs. For a deeper explanation of the PCA method refer to Refs.[82, 85]

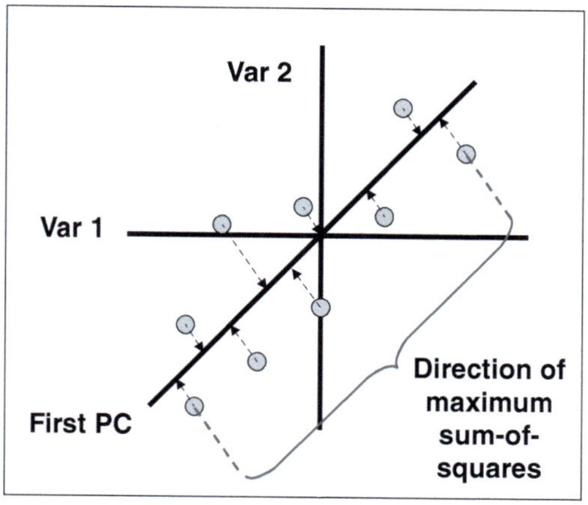

Figure 2.5 Geometrical illustration of PCA.

Example 2.2: Starting with MATLAB

Let us return to the pipe network example in order to introduce the use of PCA models in MATLAB. Through the book we will make use of either original MATLAB code or routines made by the authors. A different notation will be used to distinguish between these two types of routines. The MATLAB standard commands will be denoted as "**`standard`**" whereas the authors' own code will be denoted as ||**`authors`**||. The authors' routines, as well as the code reproducing the examples in this book, are available on Wiley's publisher website associated with this manuscript. In addition, to execute some of the examples, we will make use of the MVBatch toolbox[3],[65] which contains MATLAB code for missing batch data imputation, synchronization, calibration, and monitoring.

The pipe network data can be loaded in MATLAB with the following commands:

```
load('pipes');
X = x(:,[1 4 5]);
```

The first command loads the pipe network dataset **`x`** to the MATLAB workspace. The two-way array **`x`** contains five columns with 100 measurements each. The five columns represent the five flows which can be measured in the pipe network of Figure 2.3: $F1, F2, F3, F12$, and $F123$. With the second command, the measurements corresponding to flows $F1$, $F12$, and $F123$ are stored in **`X`**. See MATLAB help for more information regarding elements selection in a two-way or N-way array.

Let us have a look at the possible ways to obtain PCs. The PCs are the eigenvectors of matrix $\mathbf{C}_X = \mathbf{X}^T\mathbf{X}$. To compute the eigenvectors of a two-way array in MATLAB, the "**`eig`**" routine can be used:

```
[V,D] = eig(X'*X)
```

3 https://github.com/jogonmar/MVBatch/releases/tag/1.2.

Two-way array **V** contains the eigenvectors and two-way array **D** the eigenvalues in its diagonal. In the example, both two-way arrays are 3 × 3:

$$V = \begin{bmatrix} 0.4483 & 0.5734 & 0.6857 \\ -0.8217 & -0.0376 & 0.5687 \\ 0.3519 & -0.8184 & 0.4543 \end{bmatrix} \quad D = \begin{bmatrix} 1.8148 & 0 & 0 \\ 0 & 3.1390 & 0 \\ 0 & 0 & 243.1455 \end{bmatrix}$$

It can be seen that there is one eigenvalue two orders of magnitude larger than the other two. The result is that data are virtually distributed in a single direction which corresponds to the first component – third column of **V**. The distribution of the data and the first PC is shown in Figure 2.6.

An alternative approach to obtain the PCs is using the singular value decomposition (SVD) algorithm over the data array **X**:

```
[U,S,V2] = svd(X)
```

Two-way arrays **U** and **V2** contain the eigenvectors of $\mathbf{XX}^T$ and $\mathbf{X}^T\mathbf{X}$, respectively. Thus, **V2** matches the two-way array **V** obtained using "**eig**" except possibly for a sign change in the direction of each PC and the ordering of the columns. The values in the diagonal[4] of **S** are known as the singular values of **X**.

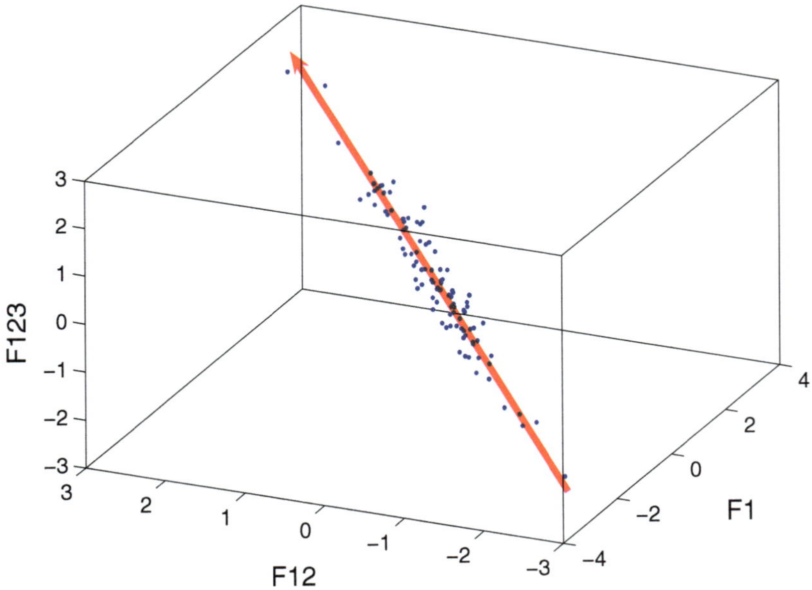

Figure 2.6 Distribution of the observations in the space spanned by variables $\{F1, F12, F123\}$ and approximate direction of the first PC. *Source*: Reproduced from Ref.[32], with permission of Elsevier.

4 The "svd" routine, by defect, yields square matrices $U(I \times I)$ and $V2(J \times J)$ and a rectangular two-way array $S(I \times J)$. The latter matrix can be simplified to a square $S(R \times R)$, where R is the rank of X, by discarding columns or rows completely composed of zeros.

The singular values correspond to the square root of the eigenvalues in the diagonal of **D**.

$$V2 = \begin{bmatrix} -0.6857 & -0.5734 & 0.4483 \\ -0.5687 & 0.0376 & -0.8217 \\ -0.4543 & 0.8184 & 0.3519 \end{bmatrix} \quad S(1:3,:) = \begin{bmatrix} 15.5931 & 0 & 0 \\ 0 & 1.7717 & 0 \\ 0 & 0 & 1.3471 \end{bmatrix}$$

Finally, another way to obtain the PCs is using the "**pca**" routine, or our equivalent ||**pca_pp**|| routine, with equivalent results:

```
[P,T] = pca(X,'Centered',false)
[P,T] = pca_pp(X,1:3)
```

The two-way array of loadings **P** contains the direction of the PCs. Thus, **P** matches **V** and **V2** except for changes in sign and the ordering of the columns. Two-way array **T** contains the scores, the projections of the observations in the latent subspace. If the sum-of-squares of the columns in **T** is computed, **sum(T.^2)**, the eigenvalues in **D** are obtained. Also, it holds that **T** is equal to the product **US**.

2.2.1 Data Preprocessing

Preprocessing generally refers to any operation over the calibration two-way array **X** done prior to model calibration, typically to mean center and/or variance normalization. Since the cost function in PCA is to maximize the sum-of-squares of the fitted data, and preprocessing alters the sum-of-squares, it has a big impact in the solution.

It is customary to center and scale the data prior to PCA. First, let us discuss the implications of centering. The trace – i.e., the sum of the elements of the diagonal – of $\mathbf{C_X}$ equals the sum of its eigenvalues λ_r:

$$\text{trace}(\mathbf{C_X}) = \sum_{r=1}^{\text{rank}(\mathbf{C}_X)} \lambda_r \tag{2.3}$$

Also, the trace of $\mathbf{C_X}$ equals the sum-of-squares of the elements of **X**. This is easy to see from the fact that the elements in the diagonal of $\mathbf{C_X} = \mathbf{X}^T\mathbf{X}$ correspond to the sum-of-squares of each of the columns. Thus:

$$\text{trace}(\mathbf{C_X}) = \sum_{j=1}^{J} \mathbf{x}_j^T \mathbf{x}_j = \sum_{j=1}^{J}\sum_{i=1}^{I} x_{i,j}^2 \tag{2.4}$$

The centering operation can be written in scalar notation as follows:

$$x_{i,j}^C = x_{i,j}^R - m_j \tag{2.5}$$

$$m_j = \frac{1}{I}\sum_{i=1}^{I} x_{i,j}^R \tag{2.6}$$

where $x_{i,j}^R$ stands for raw (non-centered) data and $x_{i,j}^C$ stands for centered data. To keep it simple at this point, we have assumed that the centering is performed across the first mode, so that the sum of the values in each column is null. Other possibilities will be discussed in Chapter 5.

If matrix $\mathbf{C_X}$ is computed from non-centered data, the following holds:

$$\sum_{r=1}^{\text{rank}(\mathbf{C_X})} \lambda_r = \sum_{j=1}^{J}\sum_{i=1}^{I}(x_{i,j}^R)^2 = \sum_{j=1}^{J}\sum_{i=1}^{I}(m_j + x_{i,j}^C)^2 \quad (2.7)$$

If the sum-of-squares of the mean elements m_j dominates the sum-of-squares of the centered data $x_{i,j}^C$, then the direction of the eigenvectors (PCs) will be determined by the former. The result is illustrated in Figure 2.7a. In the example, the first PC is roughly determined by the means of the two variables.

If, otherwise, $\mathbf{C_X}$ is computed from centered data, the PCA is focused on variance.[5]

$$\sum_{r=1}^{\text{rank}(\mathbf{C_X})} \lambda_r = \sum_{j=1}^{J}\sum_{i=1}^{I}(x_{i,j}^C)^2 \quad (2.8)$$

and a result like the one shown in Figure 2.7b is obtained. Centering is the common practice when performing PCA and the reason why it is generally stated that PCA finds the directions of maximum variance instead of the directions of maximum sum-of-squares. Centering focuses the PCA analysis on the variability within the set of observations, allowing us to determine the main patterns of difference across observations.

Now let us focus on scaling issues. In Eq. (2.8), the higher the variance of a variable, the more likely this variance will be captured by the first PCs. It could be said that the PCs are rotated toward variables of higher variance, focusing on them. Very often, variables of very different nature, measured in non-comparable units, are collected in the data array **X**. For instance, imagine a dataset in which a number of process variables collected in an industrial chemical reactor is to be analyzed. These variables include a number of temperatures and concentrations. The units of these variables, e.g., degrees Celsius and mol/liter, are not objectively comparable. If PCA is performed on these variables, the first PC would tend to rotate toward the temperatures, which are typically orders of magnitude higher than the concentrations in the units mentioned before. This is not a desirable behavior and scaling can be used to correct for these differences in magnitude among variables.

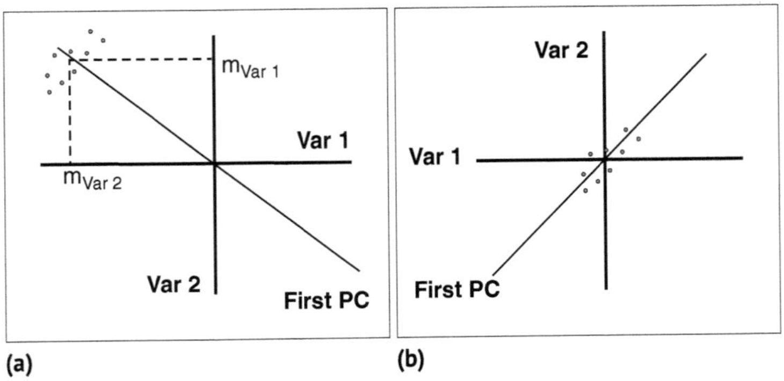

Figure 2.7 Effect of centering in PCA: (a) first PC of the raw data, and (b) first PC after mean-centering.

5 Note that the sample variance of a centered variable is proportional to its sum of squares.

Let us assume that the scaling operation is performed after centering, as it is customary. The scaling operation can be written as follows:

$$x_{i,j}^S = \frac{1}{s_j} x_{i,j}^C \tag{2.9}$$

where $x_{i,j}^S$ stands for the scaled data. Typically, centered data is scaled to column-wise unit variance, so that:

$$s_j = \sqrt{\frac{1}{I-1} \sum_{i=1}^{I} (x_{i,j}^C)^2} \tag{2.10}$$

Again, there are many alternative possibilities to this scaling method. Nonetheless, we prefer to keep it simple at this point, so that the effect of simple scaling in PCA is understood. To get a deeper insight into centering and scaling in two-way LVMs, the reading of[21] is recommended.

Centering and scaling to column-wise zero mean and unit variance is often termed auto-scaling. If $\mathbf{C_X}$ is computed from auto-scaled data, the PCA is focused on correlation:

$$\sum_{r=1}^{rank(\mathbf{C_X})} \lambda_r = \sum_{j=1}^{J} \sum_{i=1}^{I} (x_{i,j}^S)^2 = \sum_{j=1}^{J} \frac{\sum_{i=1}^{I} (x_{i,j}^C)^2}{\frac{1}{I-1} \sum_{i=1}^{I} (x_{i,j}^C)^2} = J(I-1) \tag{2.11}$$

In Eq. (2.11), all the variables contribute the same amount of sum-of-squares to the trace of $\mathbf{C_X}$, and affect equally the calibration of the PCA model. This is generally the preferred method when variables with different units are included in **X**. In Figure 2.8, the effect of scaling is illustrated. Notice that the distribution of the points around the PC have changed after the scaling operation. In Figure 2.8a, the first PC is closer to Var 1, which makes PCA to model the distribution of the observations in Var 1 with high fidelity. In Figure 2.8b, the distributions in Var 1 and 2 are equally modeled. Note the distribution of the points around the PC have changed after the scaling operation. For instance, in Figure 2.8a, the leftmost point is over the line

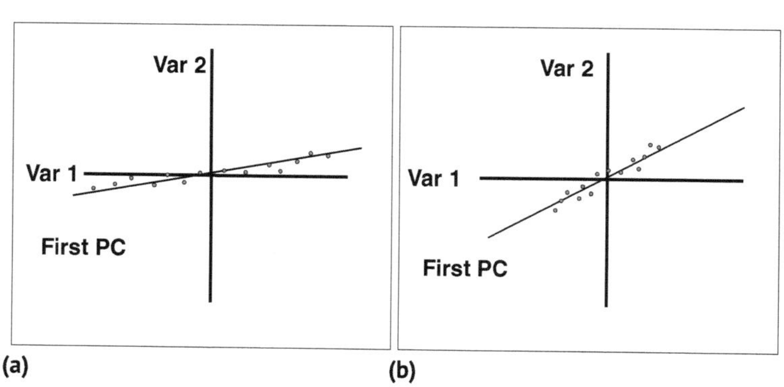

Figure 2.8 Effect of scaling in PCA: (a) first PC without scaling, and (b) first PC after auto-scaling.

representing the PC, whereas the leftmost point in Figure 2.8b is beneath that line. Although in this illustrative example, the difference between scaled and not scaled data is not dramatic, the use of different units in real data may well lead to significant differences in the distribution of the scores across PCs.

Example 2.3: The consequence of preprocessing

Let us return to the pipe network example to show how the PCA results are affected by the preprocessing step. Recall that the measurements of the flows $F1$, $F12$, and $F123$ are stored in **X**. This two-way array is mean-centered and the standard deviations of the variables are 1.1, 0.9, and 0.7, respectively. Let us create the following two new matrices of data:

```
Xm = X + (ones(size(X,1),1)*[100 0 0]);
Xs = X.*(ones(size X,1),1)*[100 1 1]);
```

The first operation introduces a mean value equal to 100 for the first variable in the two-way array **Xm**, whereas the standard deviations remain the same as in **X**. The second operation augments the standard deviation of the first variable 100 times. Thus, the two-way array **Xs** remains mean-centered but the standard deviation of the first variable is 100 times higher than in **X**. The distributions of the observations in **X**, **Xm** and **Xs** in the plane $F1$ and $F123$ are shown in Figure 2.9.

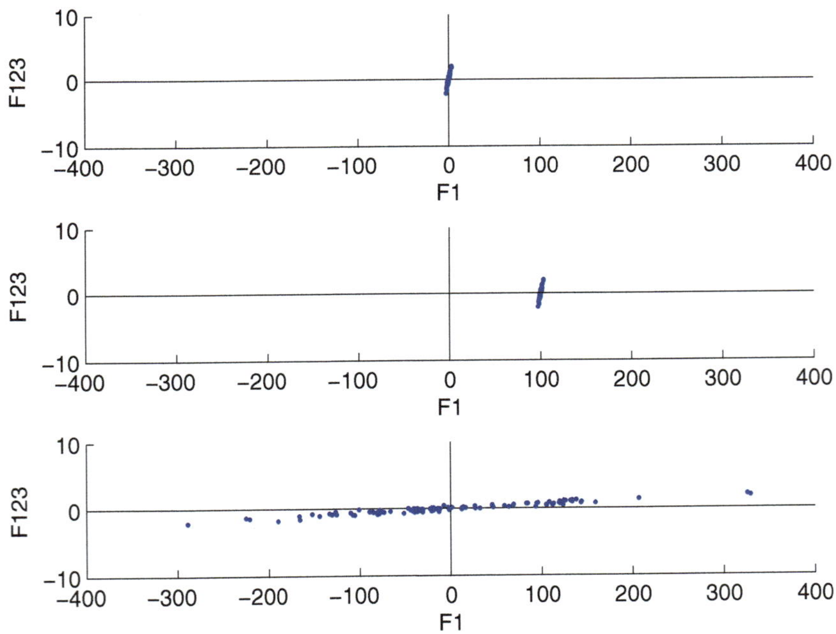

Figure 2.9 Distributions of the observations in **X** (top), **Xm** (middle), and **Xs** (bottom) in the plane $F1$ and $F123$.

Let us compute the PCs using "**eig**":

```
[V,D] = eig(X'*X)
[Vm,Dm] = eig(Xm'*Xm)
[Vs,Ds] = eig(Xs'*Xs)
```

The third column of either **V**, **Vm**, and **Vs** contains the eigenvector corresponding to the first PC:

$$V(:,3) = \begin{bmatrix} 0.6857 \\ 0.5687 \\ 0.4543 \end{bmatrix} \quad Vm(:,3) = \begin{bmatrix} -1.0000 \\ -0.0001 \\ -0.0001 \end{bmatrix} \quad Vs(:,3) = \begin{bmatrix} 0.9999 \\ 0.0081 \\ 0.0064 \end{bmatrix}$$

The first PC in **V** has similar loadings for the three variables. Nonetheless, for the altered datasets **Vm** and **Vs**, the first PC is almost completely focused on the first variable. It seems that both operations – adding a mean value equal to 100 or augmenting the standard deviation 100 times – have a similar consequence. Recall that the first variable in **V** was mean centered and presented a standard deviation close to 1. The similarity in the results for **Vm** and **Vs** comes from the fact that in both situations, the sum-of-squares of the first variable was augmented 100^2-fold. It is also a logical consequence that the first PC is focused on the first variable, since its sum-of-squares is four-orders of magnitude higher than the sum-of-squares in the other two variables.

Similar results are obtained if the PCs are calibrated using the SVD algorithm, but care should be taken when using "**pca**" because this routine automatically mean centers the dataset. If the following commands are introduced:

```
P = pca(X,'NumComponents',1)
Pm = pca(Xm,'NumComponents',1)
Ps = pca(Xs,'NumComponents',1)
```

The output is:

$$P = \begin{bmatrix} 0.6857 \\ 0.5687 \\ 0.4543 \end{bmatrix} \quad Pm = \begin{bmatrix} 0.6857 \\ 0.5687 \\ 0.4543 \end{bmatrix} \quad Ps = \begin{bmatrix} 0.9999 \\ 0.0081 \\ 0.0064 \end{bmatrix}$$

as a result of the mean centering operation inside the "**pca**" routine.[6]

Now let us introduce the command ||**preprocess2D**||. This routine can be used to center or auto-scale a two-way array. For instance, to auto-scale the two-way array **X**, the following command is used:

```
[X_as,av,sc] = preprocess2D(X,2);
```

The first output is the preprocessed data, the second one is the average and the third one is the scaling applied. Since the data were auto-scaled, **sc** contains the standard deviation. If the second argument in ||**preprocess2D**|| is 1, the two-way array is mean-centered instead of auto-scaled.

6 There is an input argument in "**pca**" that can be used to tell the routine not to mean center. The perceptive reader may remember it from our previous example.

Let us perform the same auto-scaling operation to **X**, **Xm**, and **Xs**. It can be seen that if "**eig**" is now applied to the three auto-scaled matrices, the resulting eigenvectors and eigenvalues match in the three cases:

$$\mathrm{V} = \mathrm{Vm} = \mathrm{Vs} = \begin{bmatrix} 0.6354 & 0.5121 & 0.5779 \\ -0.7614 & 0.2910 & 0.5793 \\ 0.1285 & -0.8081 & 0.5749 \end{bmatrix}$$

2.2.2 Selection of the Number of Principal Components

Much work has been devoted to find an *"optimum"* – in some sense – or at least fit-for-purpose value for A in Eq. (2.1), i.e., the number of PCs in the PCA model. A good survey on the matter can be found in the book by Jackson.[82]

Bartlett's test[10] is a statistical test for equality of the eigenvalues. The test is performed under the null hypothesis that all the eigenvalues are equal, with the alternative hypothesis that the largest one is different from the others. If the null hypothesis is rejected, the corresponding PC is incorporated into the model, and the test is repeated for the remaining eigenvalues. Otherwise, no more PCs are added to the model. Jackson[82] states that this test should be used to compute a maximum value for A. Thus, A should be equal to or lower than the test output. Another often used method consists of adding PCs until a certain amount of variance is captured in the model. This method should only be used when prior information about the non-interesting variance is available. The SCREE plot is a graphical method where the eigenvalues are plotted in decreasing order of value, and the analyst has to find a kind of an "elbow" in the shape of the curve. Nevertheless, this "elbow" is not clear in many cases. Moreover, this is a very subjective notion which cannot be univocally defined in an automatic framework for the selection of the number of PCs. Velicer[171] developed a method based on the computation of the partial correlations among the variables and the PCs removed.

Wold[184] proposed the use of cross-validation for the determination of the number of PCs. In cross-validation, data are divided into k groups. Each time, a model is calibrated on the whole dataset except for a group. Afterward, the data from that group are predicted using the model, and a criterion of goodness of fit (CGF) is computed. This is repeated for each of the k groups, and a total CGF for a model is obtained. In PCA, the CGF is computed for the models with one PC, two PCs, three PCs, and so on. From the shape of the CGF, the optimum number of PCs is estimated. Eastment and Krzanowski[44] and Nomikos and MacGregor[119] suggested the use of cross-validation when the PCA model is going to be used for future observations, which are independent of the calibration data. This is because cross-validation allows the estimation of the prediction error expected for incoming data.

The simplest cross-validation method for PCA is the row-wise k-fold cross-validation or *rkf*. See Ref.[23] for a full description of the algorithm. In this method, the observations -rows- of the two-way data array **X** are arranged into k groups, typically at random. In each cross-validation iteration, one group is left out (herein denoted as $X_{\#}$) and a PCA model is calibrated from the remaining rows (herein denoted as X_{*}). Then, the residual error $\mathbf{R}_{\#}^{A}$ two-way array of the left-out group of observations is computed with the model. This procedure is repeated until every sample is left out once and only once. Using this approach, we can compute the matrix of prediction errors $\mathbf{R}^A$ for all the observations, and from it the prediction error sum of squares (PRESS), which is used as CFG.

The *rkf* method yields strictly decreasing curves of PRESS in terms of A. To determine the number of PCs, a threshold can be applied. Thus, if the decrease of PRESS when adding the ath PC is lower than the threshold, the PC is discarded, and therefore, the model selected contains $a - 1$ PCs. Also, the curve can be corrected with the degrees of freedom consumed.[20] In Figure 2.10, a special case of the *rkf*, when the number of groups k equals the number of observations (rows) in the two-way data array, is depicted. This is referred to as the row-wise leave-one-out cross-validation.

The *rkf* has been criticized because the PCA estimates are computed using the actual values as input,[20] so that there is dependency between actual values and estimates. As a result, the modeling error computed in *rkf* cannot be considered purely prediction error (PRESS). It is not clear whether this issue is instrumental for the ability of determining A. Nonetheless, a number of cross-validatory approaches based on missing values estimation have been derived to satisfy–to a certain degree–the independence between estimates and actual values. Among the several cross-validatory algorithms proposed,[20, 59] those by Wold[184] and Eastment and Krzanowski[44] are the most cited and influential ones. The proposal by Wold is

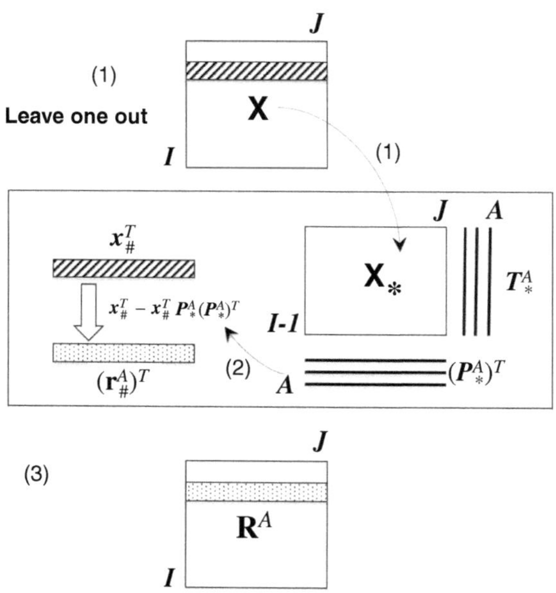

Figure 2.10 Illustration of the row-wise leave-one-out cross-validation method: (1) data is split into a calibration subset ($\mathbf{X}_*$) and a single left-out observation ($\mathbf{x}_{\#}^T$); (2) a PCA model with loadings $\mathbf{P}_*^A$ is computed from $\mathbf{X}_*$; (3) The model is used to compute the residuals $(\mathbf{r}_{\#}^A)^T$ of the left-out observation, which are stored in $\mathbf{R}_{\#}^A$.

based on the iterative computation of PCs in the NIPALS procedure.[183] Wold suggested to include PCs to the model whereas the following index is below 1:

$$R_a = \frac{\text{PRESS}_a}{\text{SSR}_{a-1}} \tag{2.12}$$

where PRESS_a is the PRESS computed for a PCs, and SSR_{a-1} is the sum of square residuals after $a - 1$ PCs have been extracted. The difference between PRESS and SSR is that the former is computed by cross-validation with missing-values estimation whereas the latter is computed at once from the entire dataset.

Eastment and Krzanowski[44] proposed an alternative scheme based on the SVD-algorithm. To select the number of significant PCs, they propose the addition of the PCs up to the last one for which the following index exceeds 1:

$$W_a = \frac{(\text{PRESS}_{a-1} - \text{PRESS}_a)/\text{DOF}_a}{\text{PRESS}_a/\text{DOF}_{\text{rem}}} \tag{2.13}$$

where DOF_a is the number of degrees of freedoms (DOFs) used to fit the ath PC and DOF_{rem} is the remaining DOFs after the ath PC has been added to the model.

Wold[184] also suggested a possible alternative with an attractive feature: the PRESS curve obtained typically shows a valley-like shape, where the minimum of the curve signals the "optimum" number of PCs. This feature is, in principle, a logical behavior for the prediction error: it decreases as the addition of PCs improves the prediction performance of the model, and it increases when this addition is noisy. It is also conveniently similar to the PRESS curve obtained in regression models. This method is referred here as the element-wise k-fold (*ekf*) method. Bro et al.[20] compared most of the principal cross-validation approaches currently used for PCA, including those introduced here, and concluded that the *ekf* method outperforms the other methods under most circumstances. It should be noted that although all methods make use of the index PRESS_a, this is computed differently in all of them.

In the forest of aforementioned methods, an important issue is often forgotten: what is the PCA model going to be used for? PCA is a very versatile tool and depending on the application at hand, the determination of the appropriate number of PCs may be addressed in a different way. From a more theoretical perspective, A is selected to maximize a given optimization function, so that it can be said to be optimum for the application at hand. Different applications imply different optimization functions, and potentially different optimum A values. Therefore, the search for a single best approach to select A for all situations seems to be ill-defined.

To determine A, we[23] propose at least three cases which should be distinguished:

- **When the interest is in the observable variables** – i.e., the variables in the two-way array $\mathbf{X}$ in Eq. (2.1). One would like to select A so that the estimation of $\mathbf{X}$ with PCA is the most accurate. The objective is by itself a definition of how A should be determined: minimizing the error between $\mathbf{X}$ and its PCA estimate. We can think of at least two types of applications within this category, depending on whether the actual values of the variables which are estimated are available or not. When the objective is data compression or dimensionality reduction,[1, 75] probably the most frequent PCA application, $\mathbf{X}$ is available and we use PCA estimates to compress the data. On the other hand, when the data in **X** is incomplete due to any problem during data collection, the actual values to estimate are not available and PCA is used to infer them.[3–5, 114, 173] In the following, these sort of applications will be termed missing data (MD) applications. Take for instance a number of temperature sensors in a chemical process. Typically, the readings of these sensors are fairly correlated and PCA can be used to develop a soft-sensor including all physical sensors. This soft-sensor may be robust to sensor failures. Thus, if one physical sensor breaks down, the missing readings can be recovered from the readings of the other physical sensors through the relationships captured by the soft-sensor. Obviously, the missing element is not used in its own estimation as it happens in compression. Notice that although the model may be built to be applied to incoming (future) data, the number of PCs is decided during model calibration with the data set at hand. In calibration, actual values are available to compute the estimation error in both compression and MD applications.

- **When the interest is in the latent variables**. This would be the case when the objective is to interpret the model to gain data understanding.[22, 32, 92] In many cases, data understanding is gained by assessing the PCA models for several numbers of PCs instead of analyzing a single model. This was the approach of Ref.[32], where cross-validation by *rkf* was combined with other sources of information in the structural and variance information (SVI) plots. Also, this is the philosophy in the missing-data method for exploratory data analysis (MEDA) approach.[22] On the other hand, there are other two-way (e.g., multivariate curve resolution) and N-way (e.g., PARAFAC) methods which are often aimed at obtaining pure components, which reflect the real underlying relationships in certain types of data. A typical example is when deconvoluting mixtures into unique and meaningful chemical compounds in analytical chemistry. In these cases, the number of components is often known *a-priori*.
- **When the interest is in the distributions in latent variables and residuals**. For most applications, the PCA model is identified solely with matrix $\mathbf{P}_A$ in Eq. (2.1). This is because it is the only matrix applied to incoming (future) objects. Nonetheless, in statistical monitoring,[47, 99] matrices $\mathbf{T}^A$ and $\mathbf{R}^A$ in Eq. (2.1) are used to develop control limits for incoming data. Therefore, $\mathbf{T}^A$, $\mathbf{R}^A$ and the control limits themselves are also part of the model of the data. This should be taken into account to select A. In this type of applications, the goal is to select A so that the statistical distributions of $\mathbf{T}^A$ and $\mathbf{R}^A$ defined from the calibration data are representative of the distributions in incoming data, provided the process under analysis remains in control. Some guidelines to select A for monitoring by assessing the stability of $\mathbf{P}^A$ with *rkf* (and therefore, indirectly, the stability of $\mathbf{T}^A$ and $\mathbf{R}^A$) were suggested in Ref.[32], but it remains as an open issue. On the other hand, the PCA model may be developed for variability reduction (Phase I) in a process using multivariate statistical process control (MSPC) techniques, which is a different objective to actual monitoring (Phase II in MSPC).[47, 163] For a more detailed explanation in Phases I and II in MSPC, the reader is referred to Chapter 7.

In this book, we are specially concerned with both process data understanding (Case b) and monitoring (Case c). Cross-validation methods based on missing data estimation, such as the approaches of Wold,[184] Eastment and Krzanowski[44] and the *ekf* algorithm,[20] are not defined for these purposes.[24] Because of this reason, the *rkf* algorithm will be generally employed throughout the book. For a more detailed investigation on PCA cross-validation methods, refer to the series of papers.[23, 24]

Example 2.4: Cross-validation

Let us illustrate the result of applying the rkf and ekf cross-validation algorithms to the pipe network example with the measurements of the flows $F1$, $F12$, and $F123$ stored in **X**. For this, routine ||**crossval_pca**|| will be used:

```
PRESSrkf = crossval_pca(X,0:3,'rkf');
PRESSekf = crossval_pca(X,0:3,'ekf');
```

The first three arguments of ||**crossval_pca**|| are, respectively, the two-way array of data, the number of PCs and the cross-validation algorithm, which can be **"rkf,"** **"ekf,"** or **"cekf"**,[140] the latter a fast approximation of the **"ekf"** approach. In the code shown, the PRESS is computed for values of A from 0 to 3. $A = 0$ stands for the model composed only of preprocessing information, e.g., the mean and the standard deviation in case of auto-scaled data. Since **X** contains three variables and 100 observations, the maximum number of components is 3. The resulting PRESS curves are shown in Figure 2.11. One PC is the recommended number of PCs for both algorithms.

2.2.3 Parameters Stability

One of the principal concerns when calibrating a PCA model should be the stability of the parameters, mainly $\mathbf{P}_A$ in Eq. (2.1) but also $\mathbf{T}_A$ and $\mathbf{E}_A$ depending on the type

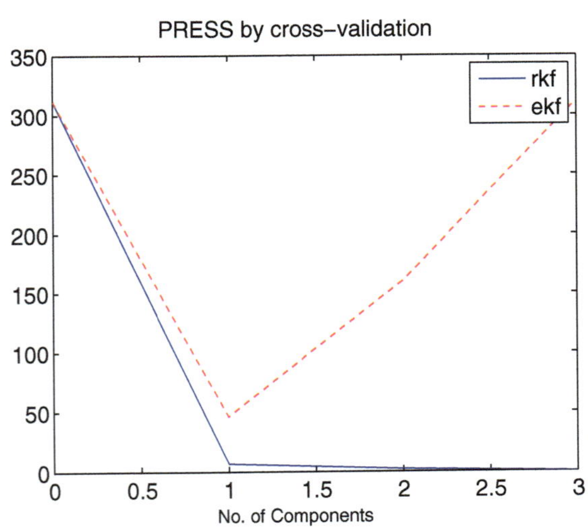

Figure 2.11 PRESS curves for rkf and ekf in the pipe network example.

of application we use PCA for. The stability – or, inversely, the uncertainty – of the parameters refers to the variance in the parameter estimates. For instance, imagine a dataset is collected in a continuous process with 100 observations of a number of process variables. This dataset is analyzed with PCA. If this process of data collection and analysis is repeated, say, five times, the resulting five PCA models will not exactly match. Nevertheless, provided PCA is an adequate technique to analyze these datasets, the trends in the variance reflected by the models should be similar.

The assessment of the parameters stability is relevant for almost any purpose PCA is applied for. For instance, if PCA is used for missing data recovery, low parameters uncertainty is desired to reduce the error of imputation. Also, if PCA is used to develop a monitoring system, low parameters uncertainty is desired to ensure a reduced amount of noise in the model. Noise affects the performance of the monitoring system, rising the number of false alarms and reducing the detection capability.

From statistics, we know that the higher the number of observations in the calibration, the better the parameter estimation, and therefore, the lower the parameter uncertainty. There is a second element which affects the uncertainty in the parameters of PCA. The more different the eigenvalues of $\mathbf{C}_\mathbf{X}$, the more stable the loadings in $\mathbf{P}_A$.[82] In Figure 2.12, we illustrate this fact with two hypothetical examples, both of them 2D: (Var 1, Var 2). We show the distribution of the noise-free observations and their counterparts with noise, and the corresponding first PC calibrated from both sets of observations. In Figure 2.12a, there is a clear direction of maximum variability, so that the eigenvalue corresponding to the first PC is fairly higher than that of the second PC. Under this scenario, the presence of noise does not affect the direction of the first PC to a large extent. In Figure 2.12b, a very different situation is shown. There is not a clear direction of maximum variability, so that the two eigenvalues are fairly similar. Thus, any content of noise may dramatically alter the identification of the first PC.

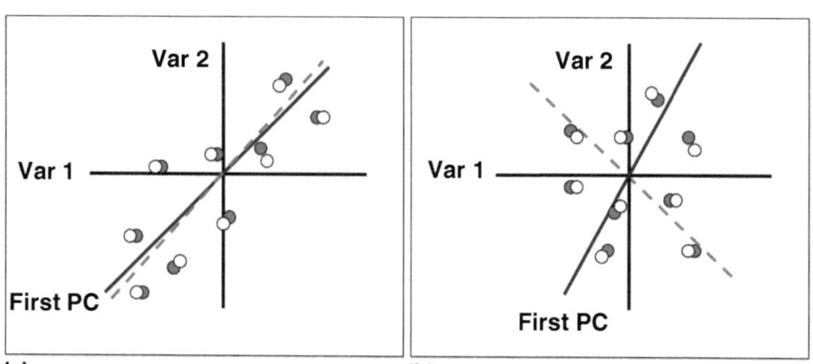

Figure 2.12 Effect of measurement noise in PCA. Noise-free and noisy observations are depicted as white and grey circles, respectively. The black and grey lines represent the true PC and the PC identified from the noisy data, respectively. In example (a), there is a clear direction of maximum variability, making the first PC quite stable. In example (b), there is no clear direction of maximum variability, causing the first PC to change significantly with small variations in the noise.

Example 2.5: Parameters stability

Let us perform a simple example to check the parameters stability in PCA. First, we create two datasets using the data of the pipe network example:

```
X1 = x(:,[1 4 5]);
X2 = x(:,1:3);
```

$F1$, $F12$, and $F123$ are stored in **X1**, and flows $F1$, $F2$, and $F3$ are stored in **X2**. Thus, whereas in **X1** there is a strong linear relationship, due to the importance of flow $F1$ in $F12$ and $F123$, the three variables in **X2** are linearly independent.

To assess the parameters stability in PCA we will compute the PCs twice by splitting each dataset into two matrices, with 50 observations each. The command "**eig**" will be used in order to obtain the eigenvalues as well:

```
[V1a,D1a] = eig(X1(1:50,:)'*X1(1:50,:));
[V1b,D1b] = eig(X1(51:100,:)'*X1(51:100,:));

[V2a,D2a] = eig(X2(1:50,:)'*X2(1:50,:));
[V2b,D2b] = eig (X2(51:100,:)'*X2(51:100,:));
```

The linear relationship in **X1** is reflected by the high eigenvalue found in comparison to the other two eigenvalues. For the first sample of 50 observations, we get **diag(D1a)' = [0.9946 1.4395 151.1776]**. Nonetheless, in **X2** the three eigenvalues are similar. Again, for the first 50 observations we obtain **diag(D2a)' = [28.9996 59.4749 78.4306]**. Therefore, the eigenvectors identified from **X1** are expected to present less uncertainty than those identified from **X2**. The eigenvectors obtained are the following:

$$\mathrm{V1a} = \begin{bmatrix} 0.4563 & 0.5516 & 0.6983 \\ -0.8230 & -0.0368 & 0.5669 \\ 0.3384 & -0.8333 & 0.4371 \end{bmatrix} \quad \mathrm{V2a} = \begin{bmatrix} 0.1438 & 0.4012 & 0.9047 \\ 0.7352 & -0.6552 & 0.1737 \\ 0.6624 & 0.6401 & -0.3891 \end{bmatrix}$$

$$\mathrm{V1b} = \begin{bmatrix} 0.4519 & 0.5955 & 0.6642 \\ -0.8207 & -0.0144 & 0.5712 \\ 0.3498 & -0.8032 & 0.4822 \end{bmatrix} \quad \mathrm{V2b} = \begin{bmatrix} 0.6086 & 0.7915 & 0.0566 \\ -0.6359 & 0.5292 & -0.5618 \\ -0.4746 & 0.3059 & 0.8253 \end{bmatrix}$$

and their differences are:

$$\mathrm{V1a} - \mathrm{V1b} = \begin{bmatrix} 0.0044 & -0.0440 & 0.0341 \\ -0.0023 & -0.0224 & -0.0044 \\ -0.0114 & -0.0301 & -0.0451 \end{bmatrix}$$

$$\mathrm{V2a} - \mathrm{V2b} = \begin{bmatrix} -0.4648 & -0.3903 & 0.8480 \\ 1.3711 & -1.1844 & 0.7355 \\ 1.1370 & 0.3342 & -1.2145 \end{bmatrix}$$

As expected, the PCs of `X1` are much less uncertain than the PCs identified from `X2`. In the latter, the uncertainty is of the same magnitude than the parameter values. Thus, these values completely depend on the set of observations considered. From the PCA of `X1` we can conclude that there is a relevant linear relationship between the three variables. From the PCA of `X2` we should not drive any conclusion regarding the relationships among the three variables.

2.3 Regression

LVMs can also be useful in regression. Unlike in compression, in regression there are two blocks of data and the aim is to estimate one block using the data in the other. The linear regression problem for two-way matrices[7] is defined according to the following expression:

$$\mathbf{Y} = \mathbf{XB} + \mathbf{F} \tag{2.14}$$

where $\mathbf{Y}$ is the $I \times M$ array of output variables to estimate, $\mathbf{X}$ is the $I \times J$ array of input variables available to estimate $\mathbf{Y}$, $\mathbf{B}$ is the $J \times M$ array of regression coefficients, the parameters of the prediction model of $\mathbf{Y}$, and $\mathbf{F}$ is the $I \times M$ array of residuals.

Ordinary least squares (OLS) performs a solution to the linear regression problem which minimizes the sum-of-squares of the residuals ($\mathbf{F}$). The least squares solution for Eq. (2.14) is:

$$\widehat{\mathbf{B}}_{LS} = (\mathbf{X}^T\mathbf{X})^{-1}\mathbf{X}^T\mathbf{Y} \tag{2.15}$$

The inversion of the matrix $\mathbf{C_X} = \mathbf{X}^T\mathbf{X}$ requires it to be non-singular. Moreover, the OLS solution is highly unstable, leading to a poor estimation performance when $\mathbf{C_X}$ is ill-conditioned. Its conditioning depends on the degree of dependence among variables in $\mathbf{X}$. That is to say, if variables of $\mathbf{X}$ are highly correlated, the estimation of $\mathbf{Y}$ should not be performed directly from $\mathbf{X}$, as shown in the following example.

Example 2.6: Estimation of variable F123 in the pipe network example by OLS

A typical application in which regression is used is in the development of soft-sensors. A soft-sensor can be very useful, for instance, in a process in which we need a more frequent sampling of a variable than what is physically plausible, or economically convenient. Since the variable cannot be measured so frequently, we need to estimate it from the other available information. Again, let us consider the pipe network example. We are collecting measurements of all the flows: $F1$, $F2$, $F3$, $F12$, and $F123$. Nonetheless, we know from the net structure that $F123$ can be estimated accurately from the other flows. Hence, we can simply estimate its value, instead of actually collecting it, saving the cost of sensor maintenance in $F123$. Additionally, the ability to estimate $F123$ can be highly beneficial if the sensor unexpectedly fails. Although the information which should be collected by the sensor will be lost until it is repaired, we may recover it from the information in the other sensors.

---

7 Like in compression, extensions to N-way matrices exist. In this book, the unfolding approach to transform three-way into two-way data arrays will be considered. Therefore, only regression of two-way arrays is discussed here. For N-way regression models, please refer to Ref.[150]

Consider the dataset with 100 observations of the five flows. We divide the data into two groups of observations, one for calibration and one to test the results of the estimation. Additionally, the calibration data is split into two in order to assess parameters stability:

```
X = x(:,1:4); Y = x(:,5);
Xcal1 = X(1:35,:); Ycal1 = Y(1:35,:);
Xcal2 = X(36:70,:); Ycal2 = Y(36:70,:);
Xtest = X(71:100,:); Ytest = Y(71:100,:);
```

Flows $F1$, $F2$, $F3$, and $F12$ are considered as predictor variables in **X** and $F123$ is introduced in **Y**. **Xcal1** and **Ycal1** and **Xcal2** and **Ycal2** are the two blocks of data of the first and second calibration dataset, respectively. Each of the calibration datasets contains 35 observations. **Xtest** and **Ytest** are the two blocks of data of the test dataset, with 30 observations.

The matrix of regressors by least squares for each of the two calibration datasets is estimated as follows:

```
B1 = inv(Xcal1'*Xcal1)*Xcal1'*Ycal1
B2 = inv(Xcal2'*Xcal2)*Xcal2'*Ycal2
```

In both cases, MATLAB outputs a warning message indicating that the results may be inaccurate since the inversions are ill-conditioned: "Warning: Matrix is close to singular or badly scaled. Results may be inaccurate. RCOND = ..."[8] The regressors matrices **B1** and **B2** we obtained are the following:

$$\text{B1} = \begin{bmatrix} 1.0840 \\ 0.3823 \\ 0.2000 \\ -0.9196 \end{bmatrix} \quad \text{B2} = \begin{bmatrix} 0.1850 \\ 0.0689 \\ 0.2000 \\ 0.6331 \end{bmatrix}$$

The results are fairly different for the two calibration data sets, especially taking into account, we have estimated only four parameters in each regressors matrix using 35 observations. To obtain the predictions for the test set, the following commands are used:

```
Ypred1 = Xtest * B1;
Ypred2 = Xtest * B2;
```

Now the true values and their estimates can be compared. Figure 2.13 compares the actual values of flow $F123$ with the predictions using the two regression matrices in the test observations. The estimation performance of the two regression models can be assessed by comparing the PRESS with the total sum of squares of **Ytest**, the latter called SSR_0:

$$\text{SSR}_0 = 12.3989 \quad \text{PRESS}_1 = 1.9640 \quad \text{PRESS}_2 = 0.1103$$

8 Two-way arrays $Xcal1' * Xcal1$ and $Xcal2' * Xcal2$ should be exactly singular, since $F12 = F1 + F2$. Nevertheless, roundoff errors in MATLAB make them only nearly singular.

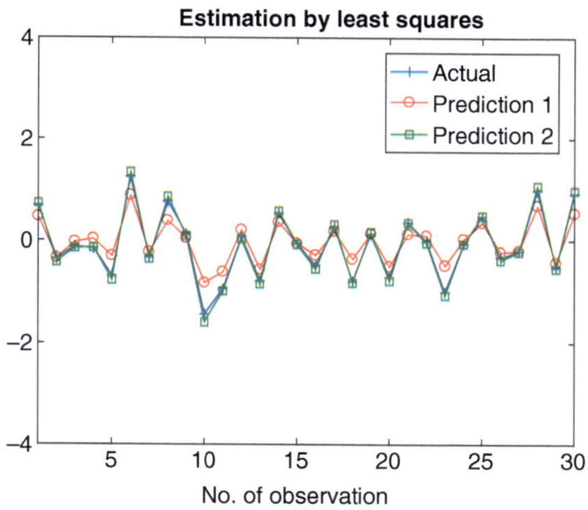

Figure 2.13 Estimates of flow $F123$ by least squares regression in the pipe network example.

For a good model, we expect the PRESS to be much lower than the SSR. While the PRESS are indeed lower than the SSR, we see a wide discrepancy between the two datasets. Recall that the pipe network example is a noise-free dataset with perfectly linear relationships among the process variables. Therefore, it should be possible to compute much better (close to zero error) prediction models. The fact that the results are far from zero error is the consequence of the ill-conditioned $\mathbf{C}_X$ inversion due to collinearity in $\mathbf{X}$.

2.4 Regression Models Based on Latent Variables

An alternative to OLS is to estimate $\mathbf{Y}$ from two-way LVMs models such as principal component regression (PCR) and partial least squares (PLS).

2.4.1 Principal Component Regression

PCR is an alternative to the least squares solution in Eq. (2.15) to the linear regression problem (Eq. 2.14) when the two-way array $\mathbf{C_X}$ is ill-conditioned or singular. PCR makes use of the PCs of $\mathbf{X}$ to estimate $\mathbf{Y}$ using least squares:

$$\widehat{\mathbf{B}}^*_{\mathrm{PCR}} = (\mathbf{T}_A^T\mathbf{T}_A)^{-1}\mathbf{T}_A^T\mathbf{Y} \tag{2.16}$$

where $\mathbf{T}_A$ is the $I \times A$ matrix of scores for the first A PCs. Since $\mathbf{T}_A$ is column-wise orthogonal, the inversion of $(\mathbf{T}_A^T\mathbf{T}_A)$ is numerically well-conditioned and the estimates of $\widehat{\mathbf{B}}^*_{\mathrm{PCR}}$ are stable provided a reasonable A is chosen. $\widehat{\mathbf{B}}^*_{\mathrm{PCR}}$ is the matrix of regression coefficients to estimate $\mathbf{Y}$ from $\mathbf{T}_A$, that is:

$$\mathbf{Y} = \mathbf{T}_A\widehat{\mathbf{B}}^*_{\mathrm{PCR}} + \mathbf{F}_A \tag{2.17}$$

Since the following equality holds:

$$\mathbf{T}_A = \mathbf{X}\mathbf{P}_A \tag{2.18}$$

with $\mathbf{P}_A$ being the $J \times A$ loading matrix corresponding to the first A PCs, the following matrix of regression coefficients to estimate $\mathbf{Y}$ directly from $\mathbf{X}$ is used:

$$\widehat{\mathbf{B}}_{\text{PCR}} = \mathbf{P}_A(\mathbf{T}_A^T\mathbf{T}_A)^{-1}\mathbf{T}_A^T\mathbf{Y} \tag{2.19}$$

Provided $\mathbf{X}^T\mathbf{X}$ is not singular, the PCR solution for A set to the number of variables is equivalent to the OLS solution. Otherwise, for $\mathbf{X}^T\mathbf{X}$ singular, A in PCR is upper-bounded by the rank of $\mathbf{X}$, and the solution by OLS cannot be computed.

2.4.2 Partial Least Squares

PLS regression is an alternative to solve the linear regression problem presented in Eq. (2.14). The basic idea of PLS is to predict a two-way array $\mathbf{Y}$ by using the latent variables or components of the two-way array $\mathbf{X}$ which maximize their covariance with the latent structure of $\mathbf{Y}$. In this case, the partial linear regression problem between the normalized two-way arrays $\mathbf{X}$ and $\mathbf{Y}$ can be defined as

$$\begin{aligned} \mathbf{X} &= \mathbf{T}_A\mathbf{P}_A^T + \mathbf{E}_A \\ \mathbf{Y} &= \mathbf{U}_A\mathbf{Q}_A^T + \mathbf{F}_A \end{aligned} \tag{2.20}$$

where $\mathbf{T}_A$ and $\mathbf{U}_A$ are $I \times A$ matrices of scores, $\mathbf{P}_A$ is a $J \times A$ matrix of loadings of the $\mathbf{X}$-block, $\mathbf{Q}_A$ is a $M \times A$ matrix of loadings of the $\mathbf{Y}$-block and, $\mathbf{E}_A$ and $\mathbf{F}_A$ are the $I \times J$ and $I \times M$ matrices of residuals of $\mathbf{X}$ and $\mathbf{Y}$, respectively. The inversion of the two-way array $\mathbf{T}_A^T\mathbf{T}_A$ is numerically stable since the latent variables in $\mathbf{T}_A$ are orthogonal and the number of components is quite lower than the number of observations. Both latent spaces are related by the inner relationship $\mathbf{U}_A = \mathbf{T}_A\mathbf{B}_A + \mathbf{H}_A$, where $\mathbf{B}_A$ is a $A \times A$ diagonal matrix and $\mathbf{H}_A$ is a $I \times A$ residual matrix. This inner relationship allows PLS to be expressed as a predictive model as:

$$\widehat{\mathbf{Y}} = \mathbf{X}\mathbf{B}_{\text{PLS}} = \mathbf{T}_A\mathbf{B}_{\mathbf{A}}\mathbf{Q}_A^T \tag{2.21}$$

where $\mathbf{T}_A = \mathbf{X}\mathbf{W}_A^* = \mathbf{X}\mathbf{W}_A(\mathbf{P}_A^T\mathbf{W}_A)^{-1}$, and $\mathbf{W}_A$ is a $J \times A$ matrix of weights. Therefore, the matrix of regression coefficients is

$$\mathbf{B}_{\mathbf{PLS}} = \mathbf{W}_A(\mathbf{P}_A^T\mathbf{W}_A)^{-1}\mathbf{B}_A\mathbf{Q}_A^T \tag{2.22}$$

Thus, a PLS model is represented by the matrices $\mathbf{P}_A$, $\mathbf{W}_A$, $\mathbf{B}_A$, and $\mathbf{Q}_A$.

Like in PCR, the maximum A in PLS is equal to the rank of $\mathbf{X}$,[9] and in this situation, both the PCR and PLS solution match. Therefore, if $\mathbf{X}^T\mathbf{X}$ is not singular, the PLS solution for A equal to the number of variables also matches the OLS solution.

Different algorithms have been proposed in the literature for PLS models, such as NIPALS[58] and SIMPLS.[39] For further details of the different algorithms and their properties, readers are referred to Refs.[77, 80, 109, 188]

9 The PLS space has up to A relevant components for prediction if and only if there are A different eigenvalues of the regressor covariance matrix.[76]

Example 2.7: Estimation of variable F123 in the pipe network example by two-way LVM

Let us use PCR and PLS to estimate flow $F123$ from the other flows in the pipe network dataset. The dataset is split into three parts, two for repeated calibration and one for testing, similar to in Example 2.7.

To obtain the matrix of regressors in PCR, the following commands are used:

```
[P1,T1] = pca(Xcal1,'NumComponents',3);
[P2,T2] = pca (Xcal2,'NumComponents',3);

B1 = P1*inv(T1'*T1)*T1' *Ycal1
B2 = P2*inv(T2'*T2)*T2'*Ycal2
```

The regressors obtained are the following:

$$\mathrm{B1} = \begin{bmatrix} 0.3810 \\ 0.0952 \\ 0.2000 \\ 0.3238 \end{bmatrix} \quad \mathrm{B2} = \begin{bmatrix} 0.3810 \\ 0.0952 \\ 0.2000 \\ 0.3238 \end{bmatrix}$$

Due to the absence of noise, the results align flawlessly. Consequently, the predicted performance is impeccable. This is shown in Figure 2.14, and it is also reflected in the PRESS of the models:

$$\mathrm{SSR}_0 = 12.3989 \quad \mathrm{PRESS}_1 = 1.4453e-030 \quad \mathrm{PRESS}_2 = 1.4473e-030$$

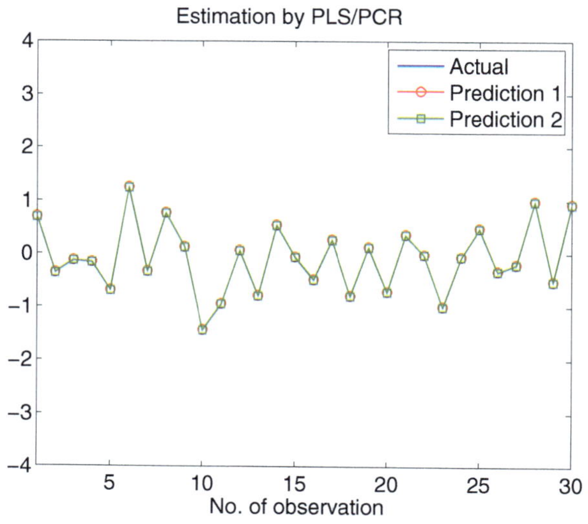

Figure 2.14 Estimates of flow $F123$ by PLS/PCR in the pipe network example. Note all profiles: actual, prediction 1, and prediction 2, have exactly the same shape and only the latter is visualized.

To obtain the two-way array of regressors in PLS, the "**plsregress**" routine is used as follows:

```
[XL,YL,XS,YS,beta] = plsregress(Xcal1,Ycal1,3);
B1 = beta(2:end)
```

```
[XL,YL,XS,YS,beta] = plsregress(Xcal1,Ycal1,3);
B2 = beta(2:end)
```

and the same results as in PCR are obtained because all possible latent variables (LVs) are used.

2.4.3 Data Preprocessing

Similarly to what happens with PCA, the preprocessing performed in the data prior to regression affects the result. The regression model typically assumes that **X** and **Y** are mean-centered, so that:

$$y^C = \sum_{j=1}^{J} x_j^C b_j^C \tag{2.23}$$

where y^C is the centered y-variable, x_j^C stand for the J centered x-variables and b_j^C are the J regressor coefficients. The model in Eq. (2.23) is equivalent to the following model:

$$y^R = y_0 + \sum_{j=1}^{J} (x_j^R - m_j) b_j^C \tag{2.24}$$

where y^R is the raw (non-centered) data of the y-variable, x_j^R the raw data of the J x-variables, y_0 is the intercept, estimated with the average of the samples collected of the y-variable, and m_j are the averages estimated for the x-variables. From both previous equations it is clear that $x_j^C = x_j^R - m_j$ and $y^C = y^R - y_0$. This is illustrated in Figure 2.15 for one x-variable and one y-variable. In this case, the model in Eq. (2.23) fits a regression line of slope b^C which passes through the average values of both variables.

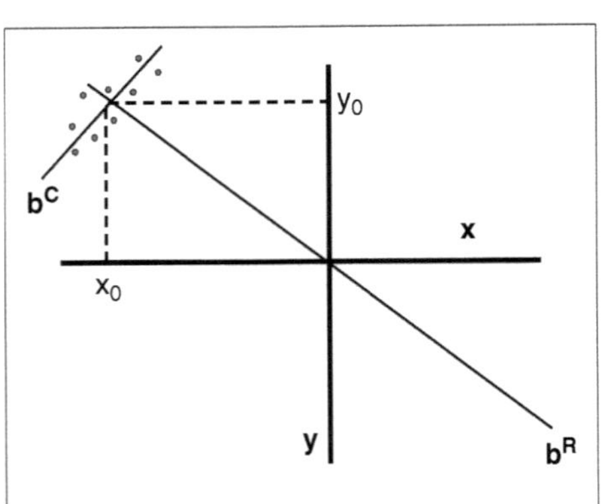

Figure 2.15 Effect of centering in regression.

Now the following question can be answered: what happens if non-centered data are used in the regression model? If variables are not mean-centered and no intercept is included in the model, the following regression model is fitted:

$$y^R = \sum_{j=1}^{J} x_j^R b_j^R \tag{2.25}$$

For the simple case shown in Figure 2.15, this model considers a regression line of slope b^R which passes through the origin. As in PCA, this regressor line will be roughly determined by the averages of x and y, provided these averages are high enough in comparison with the variability around them. Unfortunately, unlike in PCA, this situation is not corrected by adding more components in PLS or PCR. Therefore, it is strongly recommended to mean-center both **X** and **Y**. This mean-centering should only be skipped when the uncertainty of the mean, that is the parameter variance, is higher than the identified value. This is true for any of the two blocks of data (**X** and **Y**).

Regarding the scaling of the variables, the scaling within blocks does not modify the result of the regression. In other words, if all y-variables are multiplied by the same factor a_y and all the x-variables are multiplied by the same factor a_x, the result (in terms of prediction performance) remains the same, although the parameters of the model change accordingly. Nonetheless, if y-variables are scaled differently, the regression model and its performance change. This also happens with x-variables in PCR or PLS but not in OLS. The OLS prediction of **Y** is invariant to the scale of the x-variables, as it comes from

$$\widehat{\mathbf{Y}} = \mathbf{X}(\mathbf{X}^T\mathbf{X})^{-1}\mathbf{X}\mathbf{Y}, \tag{2.26}$$

so that only the scaling of y-variables influences the solution. In PCR, the scaling of x-variables is of paramount importance, since the PCs are used in the regression. As it was explained before, the PCs are rotated toward variables of high variance. Therefore, these will be the most relevant variables in the regression. In PLS, the covariance between **X** and **Y** is maximized to obtain the LVs. Therefore, the relative scaling of x-variables influences the resulting regression model, but not so much as in PCR.

Example 2.8: Preprocessing in regression

First, we will return to Example 2.7, in which the aim is to estimate $F123$ from the rest of the flows in the pipe network data set. Using a similar approach as in Example 2.3, the data is modified so that all x-variables have an average value close to 100. Also, the standard deviation of $F3$ is incremented fivefold:

```
scale = [1 1 5 1];
average = [100 100 100 100];
X = (ones (size(X,1),1)*scale .* X) + ones(size(X,1),1)*
average;
```

After this modification, `X` and `Y` are split in a block for calibration (70 observations) and a block for testing (30 observations). Now, we can compare the estimation performance when performing the regression using the modified dataset directly, after

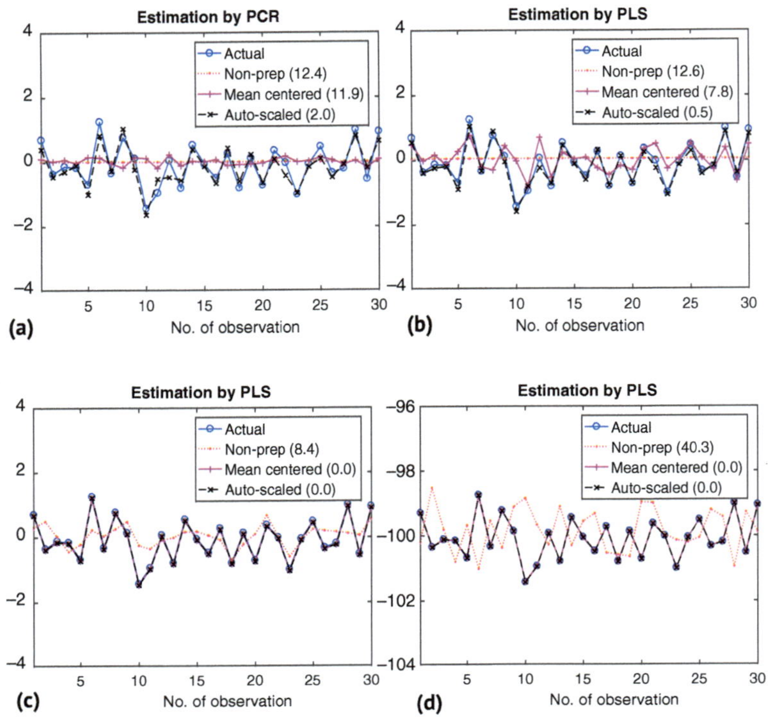

Figure 2.16 Effect of preprocessing in regression. Estimation of flow $F123$: (a) PCR with 1 LV, (b) PLS with 1 LV, (c) PLS or PCR with 3 LVs, (d) PLS or PCR with 3 LVs for an average value of $\mathbf{Y}$ equal to −100. The PRESS values of the models are shown in brackets.

mean-centering and after auto-scaling (please, note that we always consider auto-scaling in $\mathbf{Y}$). This is done in Figure 2.16. This figure compares the actual values of $\mathbf{Y}$ in the test set and the estimations using $\mathbf{X}$, $\mathbf{X}$ after mean-centering and $\mathbf{X}$ after auto-scaling. Figure 2.16a, b shows the result for 1 LV using PCR and PLS, respectively. The regression using non-preprocessed data is useless, estimating a value close to zero in all the cases. This is due to the fact that the first latent variable is roughly determined by the average value of the variables in $\mathbf{X}$. A certain improvement is obtained when mean-centering $\mathbf{X}$. Nonetheless, it is surprising that the difference in performance between the mean-centered and auto-scaled model taking into account that the increment in the variability in $F3$ was not so big. Also, differences can be observed between PCR and PLS, where the former is outperformed by the latter according to the respective PRESS values. Although it is not shown here, it can be observed that if all variables in $\mathbf{X}$ (or in $\mathbf{Y}$) are re-scaled using the same constant, the results remain unchanged. Nonetheless, if variables in $\mathbf{X}$ (or in $\mathbf{Y}$) are re-scaled using a different constant the regression model changes. This is also illustrated in Figure 2.16a, b where the mean-centered and auto-scaled data resulted in different performance outcomes.

According to the PRESS values in Figure 2.16a, b, PLS yields a better regression model than PCR for 1 LV for this noise-free example. If, instead of 1 LV, 3 LVs are computed, both PLS and PCR give the same result, shown in Figure 2.16c. The solution by both methods matches because the rank of **X** is 3, and therefore, the maximum number of LVs is extracted. Notice in this situation, mean centered and auto-scaled data yield the same result, since the complete space of the x-variables is used. Nonetheless, the regression with non-preprocessed data is still poor for 3 LVs. As explained in the previous section, the poor estimation resulting when **X** and **Y** are not centered in regression, illustrated in Figure 2.15, cannot be solved by adding LVs.

Finally, Figure 2.16d illustrates the case when the y-variable has an average value of −100. PLS with the maximum possible number of LVs (3 LVs, PCR would give the same result) is used in the regression. Like in Figure 2.16c, mean-centered and auto-scaled data (**X** and **Y**) give a perfect estimation of **Y** (remember all the experiments are noise-free). The estimation with non-preprocessed data is again poor. If there are multiple y-variables and they are re-scaled differently, the result also changes.

2.4.4 Selection of the Number of Latent Variables

The choice of the number of LVs in regression may be seen as less controversial than in compression. Authors generally agree that the way to do this is using row-wise cross-validation. As in the rkf method for PCA cross-validation, the observations -rows- are arranged in a number of groups to perform cross-validation. Typically, this arrangement is random. Also, depending on the number of rows in the groups, we have several choices.[190] In the leave-one-out cross-validation, one group per observation is considered. The leave-n-out cross-validation is a generalization of the previous approach in which we define a group per each possible combination of n observations. Thus, all observations belong to several groups. The number of groups, and so the computation time, grow fast with n, which makes this approach unfeasible in many situations. It is generally accepted that the leave-n-out approach yields the best performance, but at the expense of this high computational burden. A more computationally efficient alternative is the k-fold cross-validation, in which the observations are arranged in k groups. This approach is the PLS counterpart of the rkf method. It is a fast generalization of the leave-one-out cross-validation, the higher k the faster. Furthermore, the use of a low k value is convenient in some situations, for instance when the number of observations is high, in order to assess the parameters stability. Note that the k-fold cross-validation will be used in this book.

Example 2.9: Choice of LVs

Let us use the data in the pipe network example as follows. Flows $F1$, $F2$,and $F3$ are replicated three times in **X** with a certain amount of simulated measurement noise, where each replicate is computed as `x(:,1:3)+0.1*randn(100,3)`. Additionally, **Y** contains flows $F12$ and $F123$:

```
X = [x(:,1:3)+0.1*randn(100,3) x(:,1:3)+ ...
+ 0.1*randn(100,3) x(:,1:3)+0.1*randn(100,3)];
Y = x(:,4:5);
```

Now, the routine ||**crossval_pls**|| will be used to perform the cross-validation as follows:

```
[c,press] = crossval_pls(X,Y,0:9,7,1,1,0)
```

The first two input parameters in ||**crossval_pls**|| are the data sets **X** and **Y**. The third parameter is the number of LVs. The fourth parameter is the number of groups (k in k-fold). The fifth and sixth parameters are the preprocessing in **X** and **Y**, respectively (mean-centering was selected; 2 is for auto-scaling). Finally, we select not to have a predefined plot. The outputs are the summed PRESS for the two y-variables and the individual PRESS for each of them. PRESS curves are shown in Figure 2.17. Figure 2.17a, b shows the result for $k = 7$ and $k = 100$, respectively. It can be seen that the differences are almost negligible: PRESS curves reach a minimum for 1–3 LVs–at least, there is no clear reduction of PRESS by adding more LVs. The low error reflects that the information in the three independent variables in **X** is useful for the estimation. However, the computation for $k = 100$ took one order of magnitude more than the computation for $k = 7$.

2.4.5 PLS Versus Other Regression Models

In the literature, many alternative regression to PLS have been proposed. Some of them treat each response separately (e.g., PCR or ridge regression [RR]), while others combine multiple responses by taking advantage of the structure of the two-way array **Y** (e.g. reduced rank regression or shrunken canonical correlation models). It is worth highlighting the fact that experiments developed with PLS can be extended to PCR straightforwardly. Wentzell and Montoto have shown[176] that both methods, PCR and PLS, perform similarly, typically for PCR with more components than PLS. Other comparative studies have been performed via simulation studies to evaluate the performance of the different regression methods proposed in terms of prediction errors.[17, 50] The main conclusions from these studies are: (i) they often give fairly similar results and (ii) the results are extremely dependent on the simulation

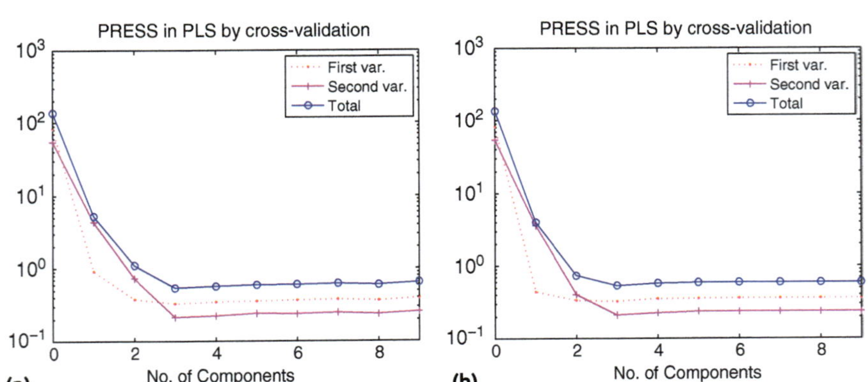

Figure 2.17 PRESS curves in PLS cross-validation shown in logarithmic scale for improved visualization: (a) $k = 7$ and (b) $k = 100$ (leave-one-out).

carried out and, thus, they may not be representative for real data found in particular fields. These comparative studies have also been criticized by several authors due to the fact that they do not take into consideration other important criteria to evaluate the performance of a regression model, such as parsimony, bias, interpretability, diagnosis for outliers and other data inhomogeneities.[17]

2.5 Multivariate Exploratory Data Analysis

Multivariate exploratory data analysis (MEDA) is an approach for data visualization and understanding based on multivariate techniques. In this section, we are interested in a dominant MEDA approach: matrix factorization with component models. PCA and PLS are variants of this type of matrix factorization. We call it matrix factorization because the original data is decomposed (factorized) into the matrix of scores, loadings and residuals. With this factorization, data visualization is simplified and enhanced. It is simplified because we can explore the distribution of the observations (rows) and of the variables (columns) of $\mathbf{X}$ in separate plots of $\mathbf{T}_A$ and $\mathbf{P}_A$, respectively. Since the number of components is typically reduced, scores or loadings are easier to visualize in bar plots and/or bidimensional or tridimensional scatter plots that retain most of the information in the data. Data visualization is also enhanced because both PCA and PLS present the main variance of interest, and therefore visualizations are typically meaningful. Another advantage is that, even if a model is created to respond to a specific question (e.g., anomaly detection), the interaction of the analyst with the data through the model can bring much more information, like the derivation of new, unexpected findings.

The plots of $\mathbf{T}_A$ are called score plots, while the plots of $\mathbf{P}_A$ are called loading plots. Clusters, trends or outliers can be identified in the plots. We can also combine scores and loadings in a single plot, commonly called a *biplot*.[54] Well-designed biplots allow to establish the interaction between observations and variables. If one observation is located far from the origin of coordinates and close to a variable in the biplot, we expect this observation to have a high value (load) of that variable. This property is useful to draw connections between the patterns of observations and variables: e.g., to identify in which variables an outlier significantly differs from the rest of observations.

Example 2.10: Illustration of MEDA

Let us illustrate the capability of PCA for data interpretation with a very simple example: The Alcohol dataset in the PLS-Toolbox (Newsweek, 127(4), 52, 1/22/1996), shown in Table 2.1. The observations (10) correspond to countries, and the variables (5) include alcohol consumption and health information. This dataset was retrieved to look for patterns between drinking habits and health variables across the countries.

Table 2.1 The Alcohol dataset in the PLS-Toolbox.[181]

Country	Liquor	Wine	Beer	LifeEx	HearthD
France	2.5	63.5	40.1	78	61.1
Italy	0.9	58	25.1	78	94.1
Switz	1.7	46	65	78	106.4
Austra	1.2	15.7	102.1	78	173
Brit	1.5	12.2	100	77	199.7
USA.	2	8.9	87.8	76	176
Russia	3.8	2.7	17.1	69	373.6
Czech	1	1.7	140	73	283.7
Japan	2.1	1	55	79	34.7
Mexico	0.8	0.2	50.4	73	36.4

We decompose the two-way data array into scores and loadings with two PCs, using ||**`scores_pca`**|| and ||**`loadings_pca`**||:

```
scores_pca(x,1:2,[],2,[],obs_l);
loadings_pca(x,1:2,2,[],var_l);
```

where we select auto-scaling (fourth and third parameters in the previous calls, respectively) and we introduce the labels of both rows (observations) and columns (variables) to provide some context in the plots. The first two PCs represent 78% of the variance of the data. The variance is a measure of the patterns of change within the data. This means that the first 2 PCs contain almost 4/5 of the total patterns of change, and we can display all this information in a reduced number of plots.

The first plot (Figure 2.18a) is a score plot, which shows the distribution of the countries. In this plot we observe two patterns, highlighted with arrows. Most countries are displayed in a quasi-linear pattern from France to the Czech Republic. As second pattern, Russia is far from the rest of the countries, manifesting a singular content in its variables in comparison to other countries.

The loading plot (Figure 2.18b) shows how variables distribute in the first two PCs. PC 1, in the abscissa, is mainly modeling the negative relationship between life expectancy and heart disease. The types of alcohol are evenly distributed in the plot, forming an almost perfect equilateral triangle. This shows that a preference for one type of alcohol leads to a reduced consumption of the other two.

Combining both scores and loadings in a biplot in Figure 2.18c, we can infer inter-connections between countries and variables. The trend from France to the Czech Republic shows where countries lie in their preference between wine and beer. Given that this trend is slightly leaning toward the right, the pattern suggests that the wine preference correlates in the data with a higher life expectancy. Finally, Russia is separated from the rest due to its preference for liqueur and higher incidence of heart attacks.

These conclusions account for 78% of the patterns of change in the data. This means that there is still 22% of change/information we have not observed yet. Most often, when subsequent PCs do not contain relevant patterns, we look at the

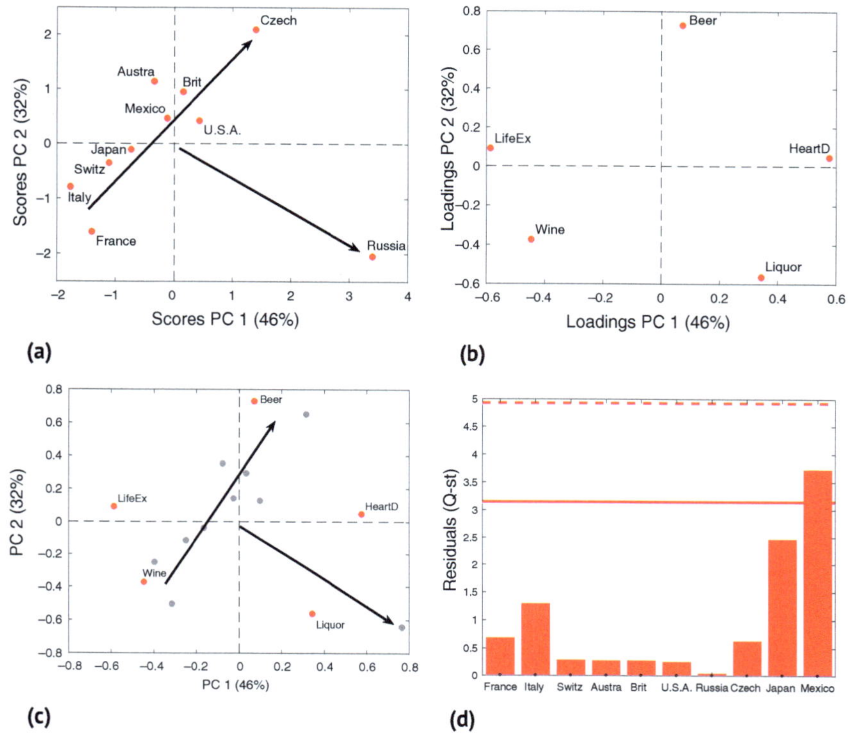

Figure 2.18 PCA plots for the Alcohol dataset: (a) score plot, (b) loading plot, (c) biplot, and (d) residuals. Q-statistics and control limits at 95% and 99% confidence intervals shown in red solid and dashed lines, respectively.

residuals as squared aggregates. For instance, the residuals can be observed as sum-of-squares in the observations (or variables). For that, we use ||**mspc_pca**||:

```
mspc_pca(x,1:2,[],2,11,obs_l);
```

where the fifth input parameter contains plotting options (type **help mspc_pca** to obtain more information about input and output parameters).

The residual plot in Figure 2.18d shows there is more to understand about Mexico and Japan than what we observed in the first two PCs. Looking at the third PC (not shown), we can infer that mainly Mexico but also Japan have lower values in all variables, a pattern that could not be seen in the first two PCs.

The matrix factorization in PCA or PLS[10] can be extremely useful to understand datasets of high dimensionality, with up to thousands of variables or even more. Matrix factorization, due to its reduced computational burden, can also be integrated into processing pipelines in which we need to interact with the data: we can easily create a new model to study in detail any pattern we find, or we can discard the data in a pattern in order to find new, more subtle, patterns.

10 The same routines showcased for PCA in the example can be found for PLS in the software MVBatch used for the examples in this book.

2.6 Missing Data

Missing data causes incomplete observations due to failures in sensors or similar problems,[6] which is a ubiquitous problem in data science that needs to be properly handled to perform a meaningful analysis of the data. See an illustration of this problem in Figure 2.19. A widespread approach to handle missing data is to discard observations/variables with a large number of missing elements. This solution is at best suboptimal, since it does not allow to make the most of all the available data. Another widely used method is to substitute each missing element with the average value of the corresponding variable computed from the data that is available. This is generally referred to as unconditional mean replacement (UMR). Either of the previous methods, dropping data or UMR, may have an impact on the correlation structure of the data, and thus on our analysis. For this reason, we introduce in this section several missing data estimation (also known as imputation) methods that take advantage of the multivariate structure. We have to recognize, however, that there is no perfect method for any possible situation, and assessing the impact of the imputation of missing data in the final model is always a good practice. This can be done, for instance, by comparing several estimates obtained with different methods.

There are a number of approaches for the use of LVMs when data include missing values. For a review, please refer to the works by Nelson et al.[114] and Arteaga and Ferrer.[4–6] Generally speaking, these methods provide an estimation of the scores of an LVM when the observations are incomplete. In turn, the estimated scores and the LVM can be used to impute the missing values.

From the point of view of the usage of an LVM, missing data can be found at two stages: model building and model exploitation. We say that missing values are found during model building when the calibration data for the LVM is incomplete. We say that missing values are found during model exploitation when an LVM is already available and the new observations are incomplete. In the following sections, the management of missing data at these two stages is discussed starting from the second one, where the model is supposed to be available.

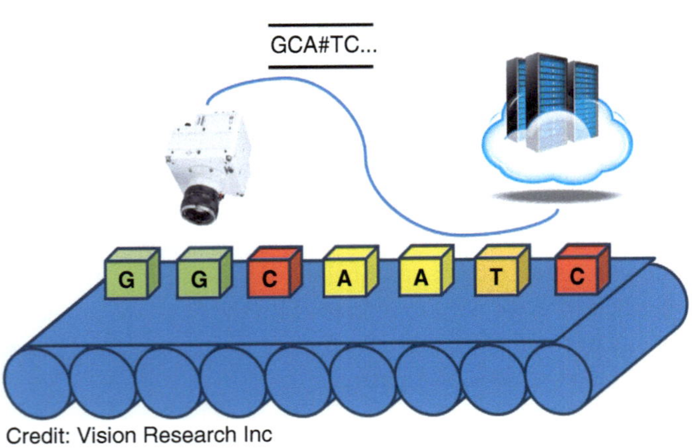

Figure 2.19 Illustration of the missing data problem. The sensor measures the blocks on the conveyor, emulating a DNA sequencing pipeline, and sends the information to a network storage system. The measurement of one of the blocks (the first A block) is missing.

2.6.1 Model Exploitation

Missing data imputation methods based on PCA models can be classified into two groups: non-regression-based and regression-based methods.[5] Relevant methods of the first group are direct estimation, also referred as the trimmed score imputation (TRI) method,[4] and iterative estimation. Both methods need an initial estimate for each missing element. This initial estimate is often computed by UMR, and its accuracy affects the imputation performance of TRI. Contrarily, in most situations, the iterative estimation is independent of the initial estimates.[23]

Figure 2.20 illustrates the geometry of both TRI and the iterative imputation methods. For simplicity, the case for two original variables and one PC is presented. In the example, two observations with the same value in variable 2 but very different values in variable 1 are shown. Let us assume the value corresponding to variable 1 in both observations gets missing. It is interesting to note that the estimates for the missing Variable 1 in the two observations will have the same value, since these estimates are obtained from the common value of Variable 2 and the same PCA model. Assume now that Variable 1 is initially set to zero. Since common data are centered previous to PCA, this is equivalent to estimating the missing element with the average of a variable (UMR). With this initial estimate, the two original observations are transformed into the point represented by the square. Subsequently, this point is projected on the PC to give the estimate of the score by TRI: the trimmed score. The trimmed score is then projected on variable 1 to obtain the TRI estimate of the original observations, represented by the circle. If this operation is repeated successively till convergence, the iterative estimate is found. The iterative estimate is represented by the triangle.

Regression-based estimation methods perform additional regression steps on the data. This results in performance improvements in comparison to direct or iterative estimation.[4] García et al.[56] present a comparative study of several estimation methods. In that paper, trimmed scores regression (TSR) along with conditional mean replacement (CMR) are recommended. In Ref.[5], the authors demonstrate that the different CMR approximations and TSR are particular cases of a general

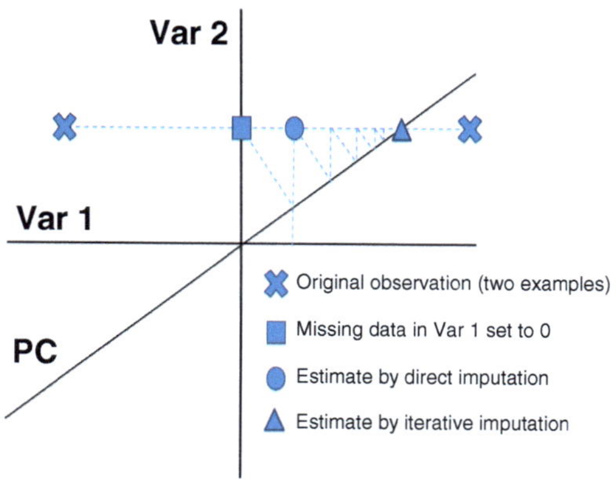

Figure 2.20 Geometric illustration of TRI and iterative imputation with one PC of two original samples in a 2D space, assuming Var 1 is missing.

framework and present similar features, where TSR is the simplest method from the algorithmic point of view. For this reason, TSR will be the preferred imputation mechanism in this book.

The TSR method[4] estimates the value of the scores from the trimmed scores, i.e., the scores obtained by TRI. Initially, the missing values of an incomplete observation $\mathbf{x}_{\text{inc}}$ are filled with zeros. Without loss of generality, let us rearrange the variables collected in the observation so that all the missing elements are located at the end:

$$\mathbf{x}_{\text{inc}} = [x_1, \ldots, x_k, 0, \ldots, 0]^T \tag{2.27}$$

The trimmed scores of $\mathbf{x}_{\text{inc}}$ are calculated directly using the PCA model:

$$\boldsymbol{\tau}_A^* = \mathbf{P}_A^T \mathbf{x}_{\text{inc}} \tag{2.28}$$

This can be re-expressed in an equivalent form:

$$\boldsymbol{\tau}_A^* = (\mathbf{P}_{A,k}^*)^T \mathbf{x}_{\text{inc}}^* \tag{2.29}$$

where:

$$\mathbf{P}_{A,k}^* = \begin{bmatrix} p_{1,1}, \ldots, p_{A,1} \\ \ldots \\ p_{1,k}, \ldots, p_{A,k} \end{bmatrix} \tag{2.30}$$

$$\mathbf{x}_{inc}^* = [x_1, \ldots, x_k]^T \tag{2.31}$$

and $p_{a,j}$ is the loading corresponding to the jth variable in the ath PC. Thus, in Eq. (2.28) the observation is filled with zeros and the complete matrix $\mathbf{P}_A$ is used. In Eq. (2.29) only the variables available in $\mathbf{x}_{\text{inc}}$ and their corresponding loadings are used.

We can make the most of the calibration data $\mathbf{X}$, which we assume on this subsection that is free of missing data, to improve the estimate of the missing elements. For this purpose, we compute a regression model that links trimmed with true scores. Let us call $\mathbf{X}^*$ the submatrix of $\mathbf{X}$ with the variables available in $\mathbf{x}_{\text{inc}}$. Then, the array of trimmed scores corresponding to the calibration data can be computed as follows:

$$\mathbf{T}_A^* = \mathbf{X}^* \mathbf{P}_A^* \tag{2.32}$$

Now, the complete two-way array of scores $\mathbf{T}_A$ can be regressed on the trimmed scores $\mathbf{T}_A^*$, so that:

$$\mathbf{T}_A = \mathbf{T}_A^* \mathbf{B} + \mathbf{F} \tag{2.33}$$

where the two-way array of regression coefficients $\mathbf{B}$ may be computed from least squares, since typically the inversion of $(\mathbf{T}_A^*)^T \mathbf{T}_A^*$ will be nicely conditioned. In any case, if it is not, PLS, PCR, or other biased methods can be used to estimate $\mathbf{B}$. Afterward, $\mathbf{B}$ is used to improve the estimation of the score in Eq. (2.29):

$$\boldsymbol{\tau}_A^{\text{TSR}} = (\mathbf{P}_A^* \mathbf{B})^T \mathbf{x}_{\text{inc}}^* \tag{2.34}$$

Finally, the score $\mathbf{t}_A^{\mathrm{TSR}}$ can be used if desired to estimate the incomplete observation, including its missing elements:

$$\widehat{\mathbf{x}}_{\mathrm{inc}} = \mathbf{P}_A \boldsymbol{\tau}_A^{\mathrm{TSR}} \tag{2.35}$$

TSR recreates the situation we have in the incomplete observation with the calibration data, and makes the most of the experience gained with these data regarding the relationship between the trimmed scores and the actual scores.

We can easily extend TSR to PLS. In PLS, the trimmed scores of $\mathbf{x}_{\mathrm{inc}}$ in Eq. (2.31) are calculated as follows:

$$\boldsymbol{\tau}_A^* = (\mathbf{R}_A^*)^T \mathbf{x}_{\mathrm{inc}}^* \tag{2.36}$$

for

$$\mathbf{R}_A^* = \mathbf{W}_{A,k}^* (\mathbf{P}_A^T \mathbf{W}_A)^{-1} \tag{2.37}$$

Recall that $\mathbf{W}_A$ is the two-way array of weights and $\mathbf{P}_A$ the two-way array of loadings corresponding with the PLS model with A latent variables. Also, $\mathbf{W}_A^*$ follows:

$$\mathbf{W}_{A,k}^* = \begin{bmatrix} w_{1,1}, \ldots, w_{A,1} \\ \ldots \\ w_{1,k}, \ldots, w_{A,k} \end{bmatrix} \tag{2.38}$$

where $w_{a,j}$ is the weight corresponding to the jth variable in the ath LV. The complete matrices $\mathbf{P}_{1:A}$ and $\mathbf{W}_{1:A}$ are used in the inversion in Eq. (2.37). Hence, the information of the complete model is used, improving the prediction capability, while avoiding problems of invertibility.

Like in PCA, the estimates of TSR in PLS are obtained by regressing the complete matrix of scores corresponding to the calibration data on the matrix of trimmed scores. Equations for the estimation by least squares of the regression parameters of TSR in PLS are provided in Ref.[29]

Example 2.11: Missing values at PCA model exploitation

Let us return to the pipe network example. The measurements of the flows $F1$, $F12$, and $F123$ are collected and measurement noise is introduced. Data are split into 70 observations for calibration, stored in **X**, and 30 observations for testing, stored in **Xmd**. Missing values are randomly introduced in **Xmd**. These values will be recovered using a PCA model with 1 PC obtained from **X**. For this data recovery, the direct, iterative and TSR imputation methods are applied using the routine ||**miss2D**||:

```
[Xmd_rec1(i,:),t1(i,:)] = miss2D(X,Xmd(i,:),1,'dir',2);
[Xmd_rec2(i,:),t2(i,:)] = miss2D(X,Xmd(i,:),1,'iter',2);
[Xmd_rec3(i,:),t3(i,:)] = miss2D(X,Xmd(i,:),1,'tsr',2);
```

The first parameter in ||**miss2D**|| is the calibration data for the PCA model. The second parameter is one – complete or incomplete – observation. ||**miss2D**|| is called into a *for* loop to recover the missing data in all the observations of **Xmd**, one per loop iteration -||**miss2D**|| only accepts one observation per call. The third parameter is the number of PCs. The fourth parameter is the missing data recovery method: **"dir"** for direct, **"iter"** for iterative and **"tsr"** for TSR. The last parameter is the preprocessing method (**2** for auto-scaling). ||**miss2D**|| returns the data with the variables estimates and the estimated scores.

The results are graphically shown in Figure 2.21. All methods perform well, since there is a nice linear relationship between the variables. In this example, the iterative and TSR methods outperform the direct method. It has been shown[24] that the iterative method is very sensitive to noise, which may lead to poor performance, whereas the TSR method is less sensitive to noise.

Now, let us assume that the estimates of the missing data in **Xmd** are treated like actual measured data, rather than estimated, which is most often what analysts do after missing data imputation. These data are projected onto the same PCA model and residuals are computed. In Figure 2.22, residuals of the datasets resulting from the direct method, the iterative method and the TSR method are shown together with the residuals from the original data. For each case, the residuals corresponding to the three variables have been rearranged (vectorized) in a vector with the elements corresponding to the missing values at the beginning. Figure 2.22 shows that

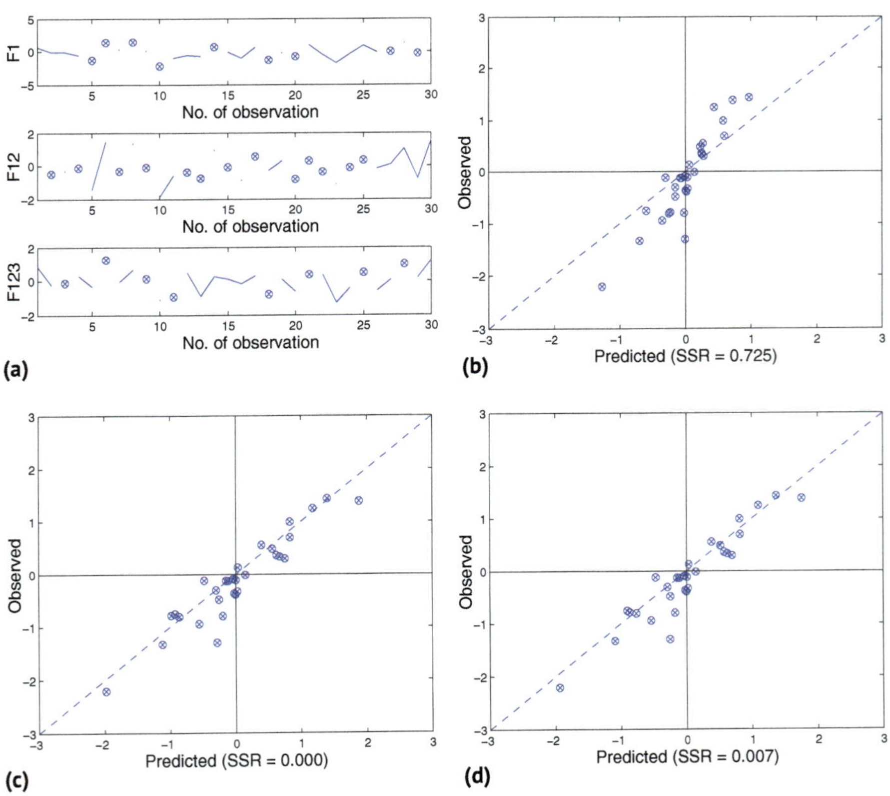

Figure 2.21 Missing data recovery in the noise-free pipe network example: (a) missing data, (b) direct method, (c) iterativemethod, and (d) TSR method. Missing data elements are marked in (a) with ⊗ symbols.

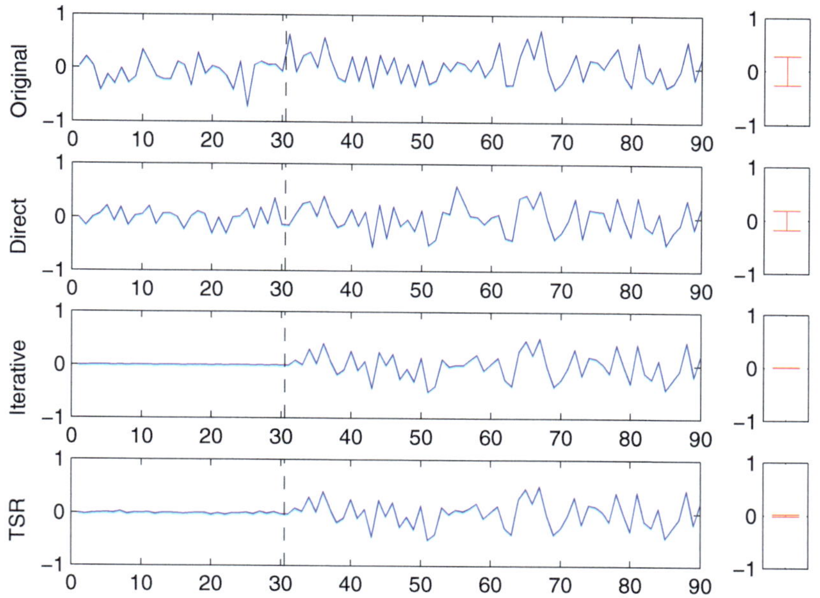

Figure 2.22 Reordered test data corresponding to the three variables in the example. Original and recovered values corresponding to the missing values in the example are shown on the left of the vertical dashed. The rest of the values are shown on the right of that line. On the right end, intervals with ±2 times the estimated standard deviation corresponding to the missing elements are shown.

the estimates present low residuals for the elements which were missing, especially those obtained with the iterative method and TSR. In the specific case of the iterative method, the residuals will always be 0 because of the iterative nature of the estimation. The iterative method converges to estimates of the missing elements so that the reconstructed observation belongs to the latent subspace.[11] Therefore, the estimated observation has no residuals (recall from Figure 2.20). In the case of the direct method or TSR, the estimates may have a residual part since the estimation is not repeated till convergence. In non-iterative approaches, the more accurate the score estimates, the lower the residuals. In any case, in PCA-based missing-data estimation approaches, the residuals in the estimates of missing values are not really "estimates of the residuals." Stating otherwise, with missing data imputation through PCA we only estimate the scores, that is, the model part of the factorization. In some applications, however, we may also want to have an approximation of the residual part through a different means.

On the right side of Figure 2.22, intervals with ±2 times the estimated standard deviation of the residuals corresponding to the missing elements are shown. The intervals computed combining actual and estimated data tend to be narrower than those computed from actual data alone (first row). This is the consequence of the absence of residuals estimation. Notice these intervals are similar in nature to

11 There are special cases where the iterative method does not converge, see Ref.[23] for more information.

the statistic used in MSPC for the residuals: the sum-of-squares. Therefore, if the amount of missing values in the calibration data or in new observations is high enough to affect the estimate of the residuals sum-of-squares, as actually happens in Figure 2.22, the performance of the MSPC system in fault detection can be seriously impacted by the imputation. This is the reason why a model of the residuals is necessary for the recovery of missing data.

2.6.2 Model Building

The discussion in the previous section was assuming that missing data were found only during model exploitation, so that calibration data was complete. When we have missing data in the calibration data, a straightforward approach is to resort to the capability of the NIPALS algorithm to build a model in the presence of a limited amount of missing elements. Nonetheless, it has been shown that when the number of missing values is high, the calibration based on TSR and the expectation-maximization (EM) algorithm yields more accurate results than NIPALS.[6] In the EM algorithm, two steps are iteratively repeated: estimation of the LVM and estimation of the missing values. These steps are repeated until convergence.

Example 2.12: Missing values at PCA model building

Let us return to Example 2.11. Imagine the data in **X** are not available. Then, only **Xmd** with missing values is available, and the missing data estimation have to be performed during model calibration. Routine ||**missTSR2D**||, which performs the EM algorithm with TSR, is useful in this situation:

```
Xmd_rec = missTSR2D(Xmd,1,2,0.7);
```

The first parameter is the calibration data matrix, which presents missing values. The second and third parameters are the number of PCs and the preprocessing method. The fourth parameter is the maximum percentage of missing values allowed in a row/column. The result is shown in Figure 2.23. Figure 2.24 shows the reduction in the intervals with ± 2 times the estimated standard deviation of the residuals due to the recovery of missing data with TSR. Although the performance of ||**missTSR2D**|| was similar to that of ||**miss2D**||, this similarity is due to the simplicity of the example. The additional information in the calibration data – **X** in the example – would be highly valuable in more complex situations. On the other hand, if we have a single dataset which presents some observations with missing values, one could be tempted to split the data into complete and incomplete observations, and use ||**miss2D**||. Nonetheless, ||**missTSR2D**|| would be more appropriate for that case, since much estimation performance can be gained from the information in incomplete observations.

2.6.3 Final Reflections About Missing Data Imputation and MSPC

Building a monitoring system for a batch process involves many steps, including data pretreatment, model building and the development of the monitoring charts. Two choices are possible to consider missing data in the design: (i) to estimate the

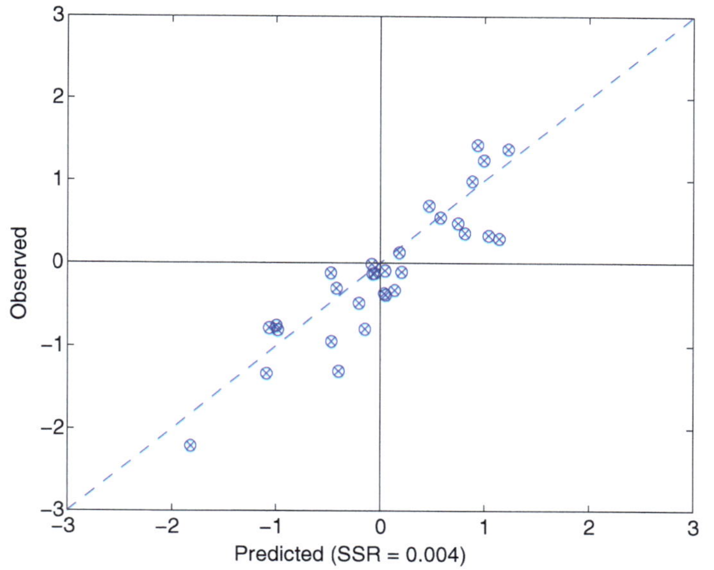

Figure 2.23 Missing data recovery using the TSR-EM method in the noise-free pipe network example.

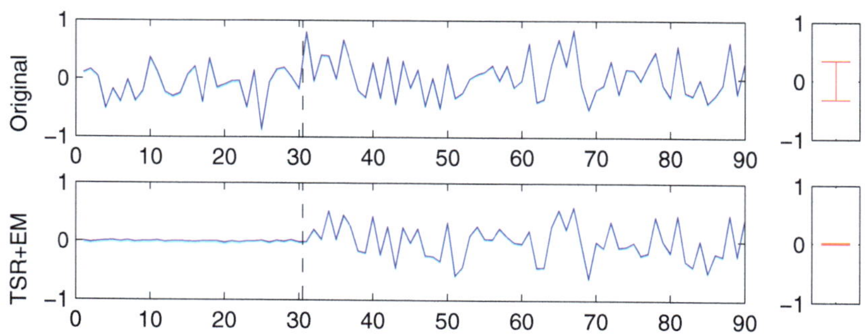

Figure 2.24 Reordered data corresponding to the three variables in the example. Original and recovered values corresponding to the missing values in the example are shown on the left of the vertical dashed. The rest of the values are shown on the right of that line. On the right end, intervals with ± 2 times the estimated standard deviation corresponding to the missing elements are shown.

missing values and treat them as actual data in the subsequent steps, as we assumed in the previous examples; or (ii) to build the monitoring system so that it explicitly deals with missing values, for instance by computing uncertainty intervals caused by the imputation. In general, choice (i) is simpler and more explored in the literature. However, we already saw that missing data methods based on LVMs give an estimate of the scores in the data, and do not provide estimates for the residuals. This may severely affect the performance of the MSPC system, which is very dependent to the accuracy of residuals estimates.

3

Batch Data Equalization

3.1 Introduction

The success of analyses based on latent variable models (LVMs) and monitoring systems relies on the quality of data collected by historian software packages.[93] Typically, process data are manipulated by compression algorithms to reduce the costs of long-term storage of manufacturing records, which is a common practice in pharmaceutical industries to achieve compliance with the demands of governmental authorities. Also, the size of databases is usually reduced when the costs of transmission of data through telecommunication networks from distanced production assets to research centers are high, as is the case with the oil and gas industry.[160] The application of univariate compression techniques to data of multivariate nature might represent a threat to the reliability of the outcomes in multivariate monitoring systems.[107]

Most industrial data historians use piecewise linear trending for data compression, although other methods based on feature extraction have been proposed in the literature.[9, 112, 175] For instance, AspenTech software uses an adaptive method based on the box-car/backward slope (BCBS) method[7] in its data historian, while OSIsoft uses a variant of the swinging door compression algorithm[18] involving a compression deviation blanket with a width equal to twice the compression deviation specification.[123] Both algorithms contain no mechanism to post-process the data in order to change the stored data point; therefore, both approaches can be considered as a filter. The main drawback of these algorithms is that the archived data can be approximated by a piecewise linear function, whereby variable values within a threshold window around the line segments are not stored. This assumption affects the way the signal is reconstructed, which is based on minimizing the deviation of the reconstructed signal from the actual signal and not aimed at preserving the variance of the signal or the correlation between the variables. The basis of these linear compression methods, jointly with the need for tuning the parameters of the algorithms, might negatively impact the trade-off between the data compression ratio and the approximation error, and hence, distort the multivariate data structure.[125]

As a consequence of compression and manufacturing reasons, measurements of process variables might be collected at different sampling times, resulting in a lack of equalization among variables, not only within a batch but also, potentially, across batches. In this scenario, the application of LVMs for the design of a monitoring system cannot be conducted because the pattern of evolution of a batch under normal operating conditions (NOC) cannot be identified. Remember that batch modeling is aimed at uncovering the effect of process variables at specific points during the evolution of the process, regardless of the time the process needs to reach these process point. The achievement of this modeling objective implies that the values of the collected process variables must be representative of the same points of evolution for all batches.[57] Hence, if batch data do not meet these requirements, the equalization of process variables must be conducted prior to the calibration of LVMs.

Since batch processes are dynamic processes that produce multivariate time-series data, it is crucial that the sampling frequency of the measured variables is greater than, or equal to, twice the maximum frequency content of the signal being sampled, which is the minimum dictated by the Nyquist–Shannon sampling theorem.[120, 144] Otherwise, the data will be aliased and fail to accurately represent the measured analogue variables. Digital Signal Processing encompasses this domain, but it also has implications for equalization policies, such as discarding intermediate values. Such issues are particularly evident when the resampling is done automatically by software without considering parameters of the data acquisition system.

3.2 Challenges in Batch Equalization

In order to model the pattern of evolution in a batch under NOC, calibration batches should be properly equalized so that they show a similar behavior at the same points of evolution during the processing.

In industrial processes, varying sampling policies can be found mostly due to three main reasons: (i) importance of some stages of the process to release products of good quality, (ii) sensor or actuator speed constraints, and (iii) the cost of high-rate sampling. If none of the previous constraints exist in the process, all process variables can be collected throughout the batch run at the same sampling rate over all batches (see Figure 3.1a). Hence, no action is required since process data are just equalized. Nevertheless, when the process variables are measured following a non-common sampling rate variable-to-variable and/or batch-to-batch, the equalization of the measurements of process variables must be performed.

A typical case of nonuniform sampling frequency is when an intermediate reactant is added to the reactor in a fed-batch chemical process, for instance, the manufacturing of specialty chemicals, such as polymers and surfactants. Typically, the feed rate or the accumulated amount of intermediate reactant added to the reactor is only monitored during this stage, whereas no value is recorded in the remaining stages (see Figure 3.1b). Note, however, that this type of processing might not represent a case of lack of equalization itself, although it has an important consequence for the way data might be arranged prior to data modeling. If some process variables are sampled at different frequencies throughout the stage and/or among batches, thereby causing a clear event of lack of equalization, data must be equalized. Otherwise, if the sampling frequency of the process variables collected only at a specific stage is the same over this period and across batches, and likewise for the rest of the variables across batches, no action is needed because data are already equalized. To continue the bilinear modeling in this case, data can be batch-wise unfolded in such a way that variables with different numbers of observations are accommodated by appending them in the columns of the unfolded two-way array,[94] as illustrated in Figure 3.1b.

Different complex sampling rates can be set up in batch processes. Usually, some stages of the manufacturing process are crucial to releasing products of good quality, and they are measured at higher sampling frequency than others (see Figure 3.1c). Take as an example the process used in Ref.[93], the emulsion polymerization. The particle size depends mainly on the nucleation period, which occurs at the first time interval of the reaction. The following process stages will also affect the particle size, but neither to the same extent as in the nucleation. Sampling more frequently than normal at this stage on all process variables may be advisable in order to monitor in detail such reaction batch to batch. In this case, the variables are not equalized but the batches are. Despite the collected data may initially appear chaotic, data can be arranged by batch-wise unfolding the three-way data array as shown in Figure 3.1c, although it is not the only possible data arrangement.

More complicated cases of sampling policy can be found in the so-called multirate systems. One of the multirate sampling cases is when process variables are sampled at different sampling rates, but this is accurately repeated in each batch (see Figure 3.1d). This phenomenon is common in sequencing batch reactors for wastewater treatment. In this kind of process, measurements belonging to pH, electric conductivity, oxidation-reduction potential, dissolved oxygen, etc., are measured at different sampling rates throughout the batch run based on their importance in the different phases of the batch, e.g., anaerobic, aerobic, and anoxic. Additionally, other measurements derived from laboratory analysis are usually obtained at a lower frequency to measure the performance of the process and the quality of the water treatment (e.g., phosphorus concentration, ammonium concentration, etc.). Hence, a set of process variables measured at multirate sampling times repeatedly over all batches is available. Another case of multirate systems is when process variables are measured following different sampling policies between variables and batches, as outlined in Figure 3.1e. This multiscale nonuniform sampling frequency is mainly caused by the application of compression algorithms by historians to reduce the size of archived data files. This type of unequalization problem may also be found in industries where the process of a plant A must be scaled up to another plant B, having different operative conditions. In these cases, it may be interesting to monitor engineering process variables, such as temperatures and pressures, at different sampling rates in phases of the process, depending on their importance in the manufacturing of on-specification products. In addition, several quality variables may be measured at lower sampling rates than the engineering variables. Due to the difficulties in obtaining desired products related to the scaling-up, this sampling policy may be different among batches in order to monitor the process more exhaustively. In the complex context of multirate systems, the use of imputation techniques is needed to

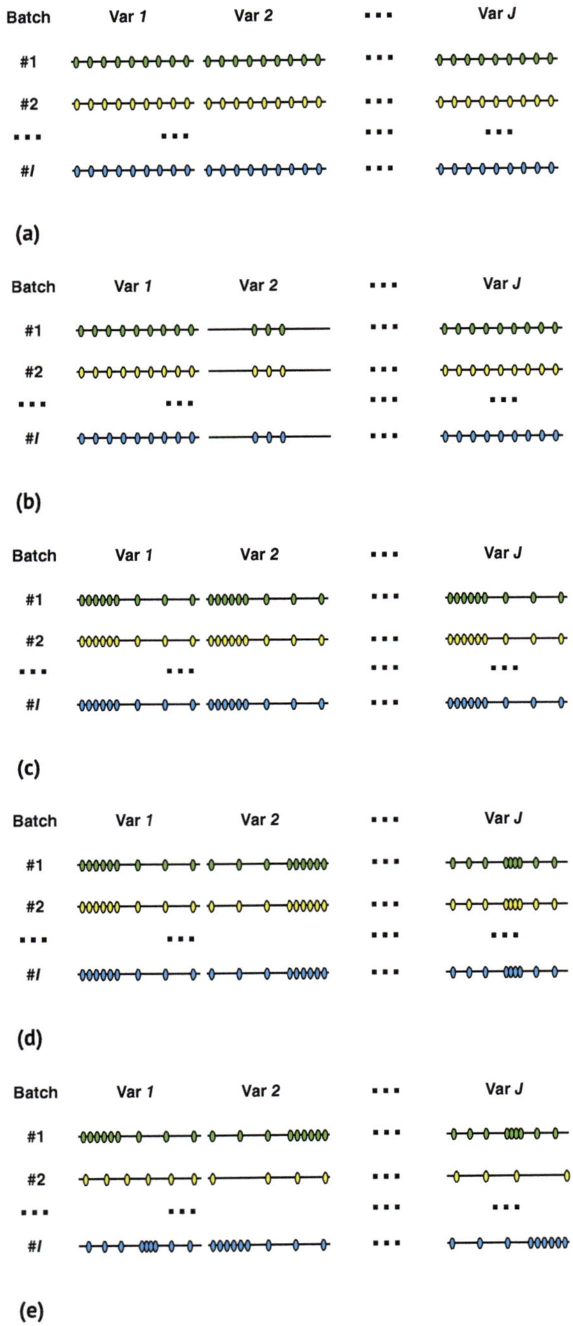

Figure 3.1 Sampling rate scenarios in batch processes: (a) same sampling rate for all variables and batches, (b) same sampling rate among variables and batches, except for a variable that is only measured at a certain phase, (c) different sampling rate in different process stages but same policy for all variables and batches, (d) multirate sampling in process variables and similar policy among batches, and (e) multirate sampling both in process variables and batches.

overcome the difference of sampling rate within and/or among batches, transforming process data to a data structure similar to the one depicted in Figure 3.1a.

For the equalization of process variables, between-batch variation is not used, but the within-batch variation is used instead because there is no certainty that batch data are synchronized, even though the batch trajectories have equal length. Hence, the time-varying correlation structure across batches cannot be modeled; only the dynamics within batches can be captured. In the following sections, we detail specific solutions tailored to address the various scenarios of unequalization encountered in batch data.

3.3 Equalization of Variables Within a Batch

A relevant step in the development of a monitoring system is how to handle a dataset in which process variables are measured at different rates, as presented in Figure 3.2a. Unfortunately, this challenge is among the most overlooked in the literature. Probably, the easiest way to build the monitoring system in multirate systems is by equalizing all the process variables to a common rate, as exemplified in Figure 3.2b. Any equalization implies the estimation of the value of one or several variables at a number of sampling times where the variables were not actually measured. In the following sections, we will assume that data acquisition cannot be modified to a sampling scheme similar to Figure 3.2b, and consequently, the use of technique for the estimation of intermediate values.

The equalization of variables can be done straightforwardly by using interpolation. In interpolation, previous and subsequent values of a variable are used to estimate its value at sampling time t according to a certain function (e.g., spline functions) that approximates data. Recall that during model exploitation in an online setup, only measurements previous to t are available for estimation.

Two principal decisions for the equalization of variables are the between-observations interval – i.e., the time elapsed between two observations – and the measurement array – i.e., the set of variables that constitute an observation. The configuration illustrated in Figure 3.2b represents the most straightforward approach. However, certain processes may benefit from an alternative data arrangement. To illustrate this idea, a simple example will be used where only one variable has a slower sampling rate than the rest. Afterward, the more general problem of multirate systems will be discussed. Take the example of a batch process aimed at the synthesis of polyhydroxyalkanoates, in which the content of biomass is measured every 7 minutes, whereas temperature, pH, and gas flows are measured every minute. To equalize the data, three options are suggested:

- **Discarding intermediate values**. Data may be reduced to a 7-minute lower frequency interval between observations so that intermediate measurements of temperature, pH, and gas flows are simply discarded from the design – for instance, collecting a sample every 7 minutes as shown in Figure 3.3a. This approach is the simplest and is desirable due to its simplicity, especially considering that this equalization will be performed online in the monitoring system. Nevertheless, this method has the drawback that any fault appearing during the

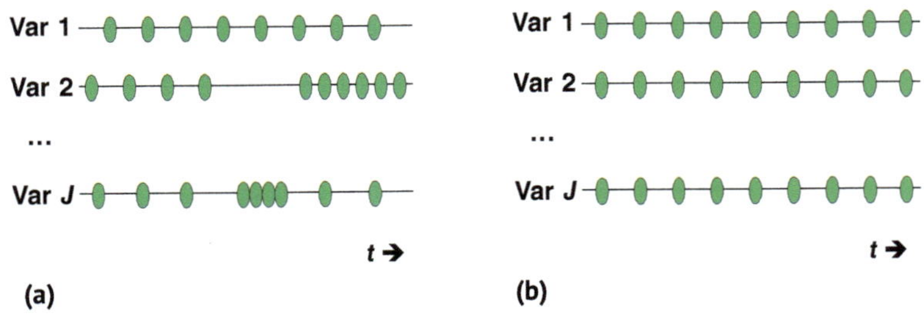

Figure 3.2 Example of (a) nonequalized variables and (b) equalized variables.

7-minute interval can only be detected at the end of this period, whereas the discarded data may contain the information to detect the fault earlier. Depending on the process, after 7 minutes have elapsed, it may be too late to fix the problem, if possible at all. Furthermore, if biomass content is measured using manual techniques instead of in-line sampling sensors, the biomass may be measured every hour or even less frequently. This is the case in many real setups. In such cases, discarding all the intermediate measurements is not feasible. It is important to note that discarding an excessive number of data points can lead to significant information loss. Therefore, it is essential to strike a balance between discarding data for equalization purposes and maintaining sufficient statistical power to detect anomalies and ensure the integrity of the information.

- **Estimating missing values.** A second choice is to estimate the value of the biomass content every minute, as proposed in Figure 3.3b. Interpolation can be used for this purpose since it might be a simple and suitable technique when the evolution of the variable trajectories – the dynamics – are smooth, although we do not exploit the multivariate nature of the data. This technique might work properly provided that measurements are auto-correlated or, otherwise stated, that the measurements are not very spaced in time compared to process dynamics. However, special caution should be taken when using univariate interpolation to ensure the integrity of the multivariate data structure. To prevent the risk of inadvertently distorting this structure, it is strongly recommended to exploit the time-varying within-batch dynamics for imputation. For this purpose, the missing data recovery ability of LVMs can be used. Note that the performance of the resulting monitoring system will clearly depend on the estimation technique.
- **Rearranging data.** A third possibility is to rearrange the data collected from temperatures, pHs, and gas flows so that all intermediate measurements are treated as different variables, as illustrated in Figure 3.3c. This option has the same between-observations interval as the first choice (7 minutes). This equalization enriches the original measurement vectors by adding lagged variables; therefore, no information is discarded as in the first option. In case a more frequent sampling rate is desired, missing data imputation techniques can be applied to estimate intermediate values, taking advantage of the process dynamics captured by adding lagged variables. The drawback of this approach is that it may lead to more complex models in terms of computational burden and interpretation.

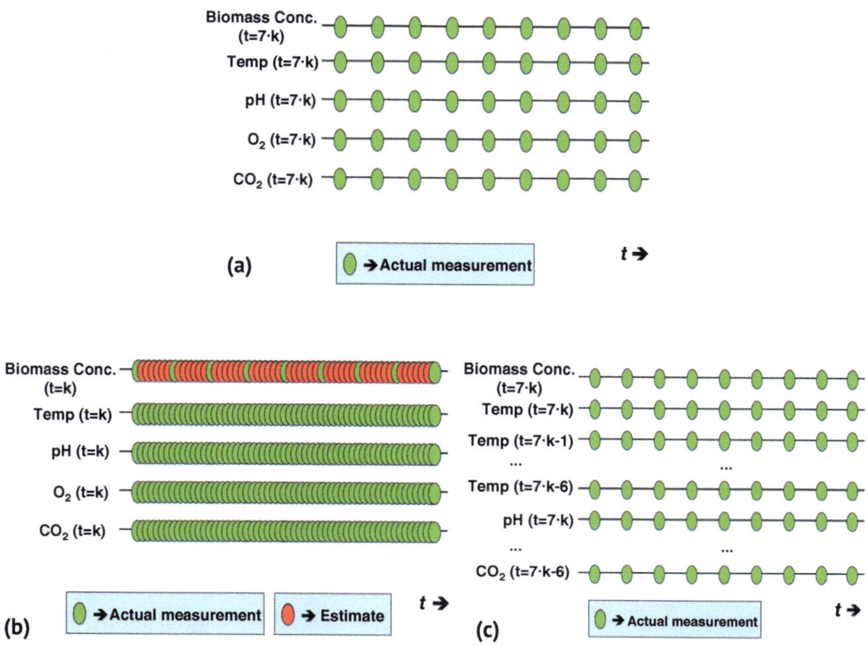

Figure 3.3 Equalization options for the fermentation process in which the biomass concentration is measured every 7 minutes whereas temperatures, pH, and gas flows are measured every minute: (a) discarding intermediate values, (b) estimating intermediate values, and (c) rearranging intermediate values.

In the preceding example, although the variables were collected at different rates, they were equalized. Thus, the biomass content was measured every seven measurements of pH and the others. Nonetheless, the example is also valid for situations when there is no equalization. For instance, this would be the case if temperatures, pH, and gas flows were measured every 2 minutes. A possible solution is to interpolate these process variable flows to a 7-minute rate.

On the other hand, option (i) in the list above is not applicable to the more complicated case of multirate systems. In this situation, all process variables may be collected at different sampling times. Therefore, hardly any complete observation may be found, and the complete dataset would be discarded according to option (i). Also, option (iii) is not applicable in this case. Option (ii) is the most generally applicable method. If the missing data algorithms inherent in LVMs are used for the estimation in option (ii), instead of univariate interpolation, the two-way data array may be enhanced with additional columns with lagged variables, similar as in option (iii), considering dynamic information for the missing data estimation. Also, multimodel approaches may be necessary. The implications of applying these solutions are studied by using batch data from the simulator of the *Saccharomyces cerevisiae* fermentation process[65] in the following subsections. For the sake of simplicity, the case of one variable being collected at a lower rate than the rest of the process variables is considered.

The comparison of the equalization techniques is conducted by comparing the LVMs fitted on within-batch data, which is the only source of accurate information available at this modeling stage. Under this premise, the multivariate models built in this section are not intended to be used for process understanding, troubleshooting, or monitoring, but rather to illustrate how the latent structures, and more precisely, variances and instantaneous cross-covariances of the process variables, for a particular batch are affected by equalization.

3.3.1 Discarding Intermediate Values

When reconstructing continuous measurements from discrete samples – i.e., continuous measurements with missing values – it is essential to consider the Nyquist–Shannon sampling theorem. This theorem states that a continuous signal can be accurately reconstructed only if it is uniformly sampled at a rate that exceeds the highest frequency of the signal by at least a factor of two. This principle is crucial to avoid a type of distortion known as aliasing or loss of information. Consequently, discarding intermediate values is feasible as long as it does not result in the loss of significant information describing the main phenomena driving the process and is relevant to the modeling objectives, such as fault detection, diagnosis, or missing data imputation. These considerations are closely aligned with the principles of the Nyquist–Shannon theorem. In the next section, the reader will be provided with an illustrative example to study the implications of discarding intermediate values as a procedure to equalize data by securing the same between-observation interval across process variables. Note that the results can be reproduced by following the set of instructions available in the MATLAB® live script "ex3_1.mlx," and summarized in Section 3.1.

Example 3.1: Discarding intermediate values as an equalization method

To initiate this example, first we need to load the raw batch data generated using the simulator of the *Saccharomyces cerevisiae* fermentation process.

```
load('saccha_30batches.mat')
```

Prior to proceeding with the equalization according to option i) discarding intermediate values, we will set the biomass concentration to one-tenth of the sampling frequency of the rest of the variables – i.e., one measurement every 10 measurements of the other variables. Note that this manipulation of the sampling frequency will be applicable to all the examples in this chapter. The resulting equalized variables are shown in Figure 3.4b. In addition to potential aliasing problems caused by the resampling, it can be seen that there is an important loss of information in the equalization. In particular, there are more abrupt changes in the shape of the equalized variable (see Figure 3.4b) than in the unequalized variables (see Figure 3.4a). Note that the less frequently a variable is sampled and subsequently equalized, the more changes in the shape of the variable trajectories. Modifications of the original variable may mask mild changes and abnormal situations, which may not be detected by the monitoring system.

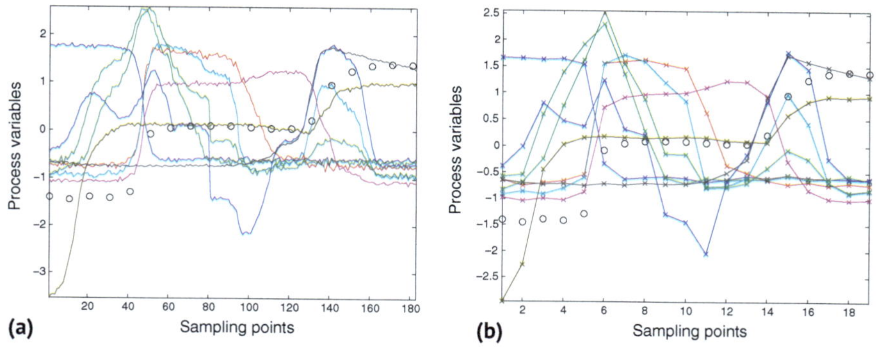

Figure 3.4 Measurements of the process fermentation, where the biomass concentration (open circles) is measured (a) every 10 measurements of the other variables and (b) equalization following option (i) discarding intermediate values.

With the objective of studying the implications of the application of this equalization method on LVMs, a PCA model is calibrated, extracting the first two principal components for the unequalized and equalized datasets after autoscaling, i.e., centering and scaling data so that each variable has variance equal to 1, respectively.

```
data_noneq_as = preprocess2D(raw_data,2);
[p,t] = pca(data_noneq_as,'NumComponents',2,'Centered',0);
data_eq_i_as = preprocess2D(data_eq_i,2);
[p_i,t_i] = pca(data_eq_i_as,'NumComponents',2,'Centered'
    ,0);
```

Figure 3.5 compares the scores, loadings, and residuals of a PCA model with two PCs calibrated from the original raw data, where biomass concentration is measured at the same sampling rate as the other variables (see Figure 3.4a), with those from the model calibrated from the equalized data obtained by discarding intermediate values after autoscaling the process variables (see Figure 3.4b). There are differences between both models, as noticeable in Figure 3.5. The score trajectories formed by the score values obtained for each sampling time of the selected batch corresponding to the original data and the data equalized by choice i) do not overlap, although differences are not very significant (see Figure 3.5a). Likewise, the positions of the loadings in the plot shown in Figure 3.5b differ between models, although the differences are not large. Looking at the residuals of the biomass concentration from both PCA models in Figure 3.5c, slight differences are observed again. In general, it is not assumable to discard a large percentage of the measured data solely to align with the least frequently sampled variable. While there is a risk of losing valuable information, in this particular case, we did not observe a significant deterioration, although it may be relevant for the performance of a fault detection system.

Unfortunately, in a real situation, the model built from the original data is not available for comparison. A possible alternative is to compare the models obtained from data equalized following options (ii) and (iii) illustrated in Figure 3.3b, c.

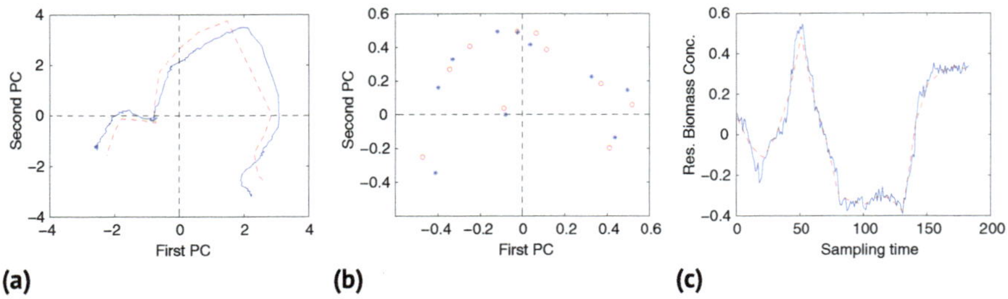

Figure 3.5 Comparison of the PCA models from the original data – no missing values – (blue lines and markers) and the data after discarding the intermediate values (red lines and markers): (a) scores, (b) loadings, and (c) residuals of the biomass concentration.

3.3.2 Estimating Missing Values

There are a number of approaches to impute missing data, from the simplest approach, such as the univariate interpolation of intermediate values, to the methods based on exploiting the correlation among variables using LVMs. The former consists of calculating new data points within a specified range of a discrete set of sampled points by using low-degree – e.g., spline interpolation – or high-degree – e.g., polynomial interpolation – polynomials in each of the intervals, choosing the polynomial pieces such that they fit smoothly together.[130] The latter are estimation methods based on PCA models, which can be classified in two groups: non-regression-based and regression-based methods.[5] For model building, the calibration based on the trimmed score regression (TSR) in an iterative manner – so-called trimmed score regression model building (TSR-MB) – is used due to its accurate results in missing data imputation.[6, 49] In the TSR-MB algorithm, the first step consists of filling the missing positions of the new observation with an initial estimate, followed by two steps that are iteratively repeated until convergence: (i) calculating the estimated vector of scores for observations using the TSR algorithm and the latent structure from the PCA model fitted, and (ii) re-estimating the missing values by reconstructing the original two-way data array with the estimated vector of the TSR-scores and the loading vector of the PCA model. For details on this imputation technique, readers are referred to the original work.[4, 49] In the following, differences between the two aforementioned approaches for estimating the missing values in the biomass concentration are studied.

Example 3.2a: Interpolation

A univariate interpolation based on cubic splines is carried out to impute the nine intermediate values between samples (see Figure 3.6a). After reducing the sampling frequency for the biomass concentration, as explained in the first example, equalize this variable using univariate interpolation based on cubic splines. This will allow us to impute the nine intermediate values between the samples.

```
data_eq_ii_a = data;
x = find(~isnan(data_eq_ii_a(:,6)));
y = data_eq_ii_a(x,6);
```

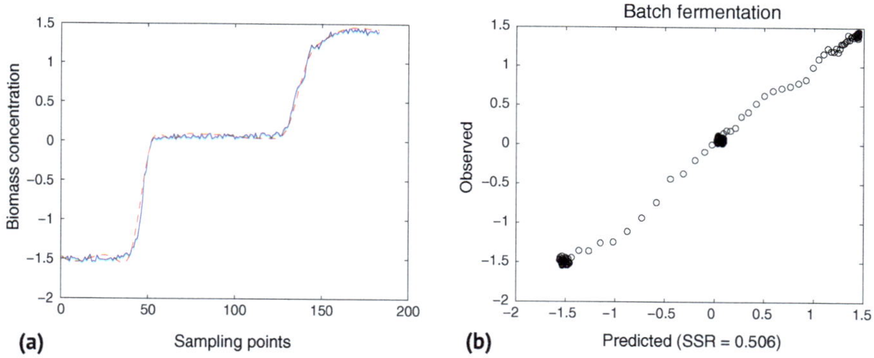

Figure 3.6 Comparison of the original data (blue line) with data resulting from the estimation of the biomass concentration using cubic spline interpolation (red line), shown in two views: (a) time series and (b) predicted versus observed values. Note that SSR is calculated as SSR $= (\mathbf{x}_j - \widehat{\mathbf{x}}_j)^2$, where $\mathbf{x}_j$ and $\widehat{\mathbf{x}}_j$ are the column vectors containing the sampled and estimated samples of the biomass concentration $j = 6$ of the selected batch.

```
data_eq_ii_a(:,6) = interp1(x,y,1:size(data_eq_ii_a,1),
    'spline');
```

Carrying on with next steps provided in the MATLAB live script “ex3_2a.mlx,” we preprocess the raw and the resulting equalized two-way data arrays using the autoscaling method, respectively.

```
data_noneqori_as = preprocess2D(raw_data,2);
data_eq_ii_a_as = preprocess2D(data_eq_ii_a,2);
```

As shown in Figure 3.6, the estimation appears accurate, since the original and equalized trajectories almost overlap; however, oscillations have been introduced in the interpolation. In Figure 3.6b the sampled and equalized values of the biomass concentration are depicted for comparison. The results of this specific simulation of a multirate sampling scenario suggest that cubic interpolation performs adequately, provided that the criteria are met, where predicted versus sampled values roughly lie on a diagonal line and the sum of squared residuals (SSR) of the biomass concentration is low (SSR $= 0.506$). However, a low SSR alone is not decisive to confirm that interpolation meets the main requirement to continue modeling batch data, but the preservation of the original time-varying covariance structure.

At this point, it is worth cautioning against the danger of applying univariate interpolation on dynamic data and the effects on the relationships of the process variables captured in the covariance matrix. The main problem polynomial interpolation might generate is the so-called Runge’s phenomenon, which is a problem of oscillation that occurs when using high-degree polynomials over a set of equi-spaced interpolation points.[138] These oscillations can be found at any interval, although they are more frequent at the edges of the intervals (i.e., start and end points), and with more severity, at the surrounding points of an outlier.[94]

An example of the oscillatory problem can be found in the interpolated variable using cubic splines, depicted in Figure 3.6a. Comparing the actual and interpolated values of the biomass concentration, there exist damped oscillations produced by the polynomial interpolation at the first stage of the fermentation – the first 40 sampling times – which is an artificial behavior not observable in the original variable (see solid blue and red dashed lines in Figure 3.6a). In addition, these oscillations are reproduced in the transition between the two exponential growth stages – between sampling times #50 and #130. The difference between this oscillation and the one at the start of the batch lies in the generated trend. As discernible from sampling time #60 to #130, a negative trend is created by the univariate polynomial interpolation as a consequence of the change in the signal induced by the second exponential growth stage. This behavior might cause a change in the time-varying dynamics and, hence, in the instantaneous and dynamic relationships of the process variables. As a result, the bilinear modeling of batch data might be severely jeopardized. Unrealistic correlations might be generated, compromising the interpretation of the statistical model, the prediction of quality properties, the imputation of missing data via statistical soft sensors, and the detection and diagnosis of abnormal occurrences within the process. Despite some authors claiming that the estimation accuracy is not the main concern when the aim is to detect faults,[31] the addition of artificial correlations might increase the false alarm rate or even mask failures in the process. A sensible way to proceed is to compare the model obtained after different equalizations and check whether the models are similar or not. If differences exist between models, the underlying reasons should be investigated. Also, it is useful to have a list of all the steps performed in the model building, since unexpected results in the monitoring system may be caused by decisions at the very beginning of the bilinear modeling cycle, such as those related to data equalization. If there are suspicions that the multivariate correlation structure among batches is compromised, the resulting equalization should be discarded.

Example 3.2b: Missing data recovery using a single-phase non-dynamic model
Another possibility tailored to batch process modeling needs is to use the ability to recover missing values of the LVMs using latent structures, which exploits the multivariate nature of the data for imputation purposes. To illustrate the effects of imputation of intermediate values using the TSR-MB algorithm, we use the same data as in previous cases, in which we have reduced the sampling frequency of the biomass concentration. The code to execute this example can be found in the MATLAB live script called "3_2b.xlm." As a first step, we equalize the biomass concentration using the TSR-MB algorithm coded in the function ||**pcambtsr**|| as follows:

```
data_eq_ii_b=pcambtsr(data,5);
```

Next, we autoscale the raw and the resulting equalized two-way data arrays, respectively.

```
data_noneqori_as = preprocess2D(raw_data,2);
data_eq_ii_b_as = preprocess2D(data_eq_ii_b,2);
```

The outcomes of applying this procedure to the dataset under study are shown in Figure 3.7. In terms of preservation of the dynamic data structure, Figure 3.7a shows that the damped oscillations created by polynomial interpolation are not present.

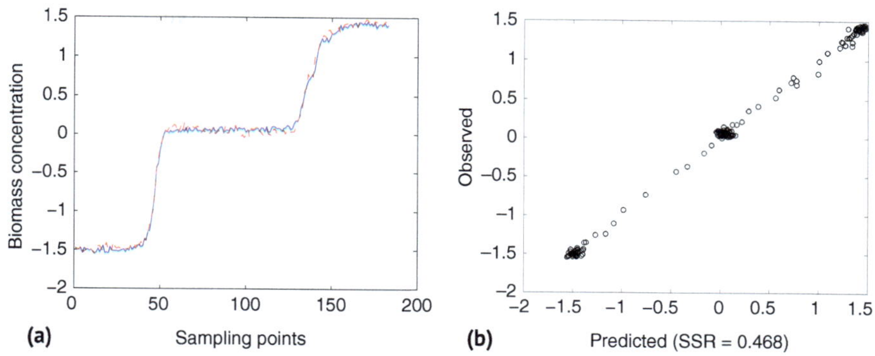

Figure 3.7 Comparison of the original data (blue line) with data resulting from the estimation of the biomass concentration using the TSR-MB algorithm (red line), shown in two views: (a) time series and (b) predicted versus observed values. Note that SSR is calculated as $\text{SSR} = (\mathbf{x}_j - \hat{\mathbf{x}}_j)^2$, where $\mathbf{x}_j$ and $\hat{\mathbf{x}}_j$ are the column vectors containing the sampled and estimated samples of the biomass concentration $j = 6$ of the selected batch.

For comparison purpose, let's calibrate a PCA model with two principal components on the previously autoscaled, unequalized, and equalized datasets, respectively.

```
[p,t] = pca(data_noneqori_as,'NumComponents',2,'Centered'
   ,0);
[p_ii_b,t_ii_b] = pca(data_eq_ii_b_as,'NumComponents',2,
   'Centered',0);
```

In Figure 3.7b, the residuals of the intermediate samples are managed in the PCA model fitted in the TSR-MB algorithm by assigning these values to the average value. When data are mean-centered in the first iteration of the TSR-MB algorithm, the algorithm assigns zero values. Note that the variable trajectories of the fermentation process are characterized by having long periods of noise with low variability (e.g., glucose, acetate, and biomass concentrations) and periods of high variability containing common cause process variation. To prevent noise from having more weight and signal from having less weight in the model, which would negatively affect the quality of the missing data imputation, the batch trajectories are not autoscaled. To overcome the problem of the different scales of the process variables, more PCs are extracted to account for those components explaining less variability. The estimation of the intermediate values using this procedure outperforms the interpolation method (SSR = 0.468 for missing data imputation by TSR-MB versus SSR = 0.506 for polynomial interpolation). This is mainly due to the high collinearity and data redundancy present in the batch trajectories, which permits accurately imputing the missing values and reducing the SSR. However, as explained above, the main difference lies in the preservation of the time-varying dynamics within the batch run when missing values are imputed, unlike when missing values are interpolated.

Example 3.2c: Missing data recovery using a single-phase dynamic model

A drawback of the estimation using a single-phase approach in the previous example is that the dynamics in the variables are not taken into account. In order to include dynamic information in the estimation is to add lagged variables as additional columns. In the next example, whose code you will find in the MATLAB live script "ex3_2c.mlx," one order of dynamics – equivalent to one lagged measurement vector– is selected. To add these dynamics, each row of the resulting matrix of data must contain the measurement of the process variables at time k and $k-1$.

By adding information of several orders, i.e., several lagged measurement vectors (LMVs), the estimates of the missing values of a variable are not only based on instantaneous values of other variables, but also on past values of the variable itself and the other variables.

Next, we proceed with the imputation of the missing values leveraging the dynamics added to the dataset, using seven PCs as the optimum number of PCs.

```
data_eq_ii_c=pcambtsr(data_lagged,7);
```

Then, we recover the original process variables by removing the lagged measurement vector.

```
data_eq_ii_c = data_eq_ii_c(:,1:size(data,2));
data_eq_ii_c = data_eq_ii_c(:,[2:6 1 7:end]);
```

Thereafter, we preprocess the modified raw and resulting equalized two-way data arrays, respectively, by autoscaling the process variables.

```
data_noneqori_as = preprocess2D(raw_data,2);
data_eq_ii_c_as = preprocess2D(data_eq_ii_c,2);
```

The result of using this approach for the whole batch using two LMVs is shown in Figure 3.8. Similar to the single-phase non-dynamic model approach, the exploitation of the multivariate data structure avoids the addition of oscillations that potentially might affect the outcomes of later steps in the modeling cycle, as observed in Figure 3.8a. The dynamic model approach slightly outperforms the non-dynamic approach, using the difference between observed and predicted values as the criterion (for dynamic versus SSR = 0.468 for non-dynamic approach). The addition of lagged variables, subsequently capturing more process dynamics, shows that the accuracy of the LVM to impute missing values can be further enhanced with information about dynamics. Hence, both methods show slightly better results in comparison to the interpolation results in terms of accuracy in prediction.

In the calibration of LVMs, we are assuming that the relationship among process variables remains the same during the evolution of the batch. However, batch processes are characterized by having time-varying relationships among process variables. In addition, the limitation of using the within-batch data hampers the removal of nonlinearities in data. Since the bilinear PCA method is used for equalization, the accuracy of the imputation may be potentially affected because not all the complex relationships can be captured. To study the accuracy of the equalization and its implications in modelling, a PCA model is calibrated with two PCs on the previous autoscaled, unequalized, and equalized datasets, respectively, similarly to the previous examples.

```
[p,t] = pca(data_noneqori_as,'NumComponents',2,'Centered'
    ,0);
```

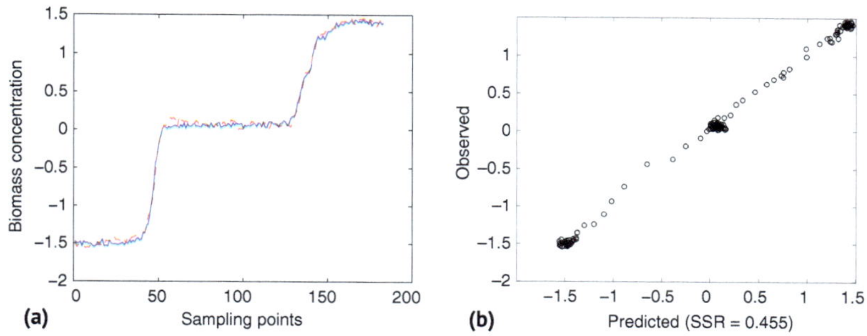

Figure 3.8 Comparison of the original data (blue line) with data resulting from the estimation of the biomass concentration using the TSR-MB algorithm in a dynamic model approach with two LMVs (red line), shown in two views: (a) time series and (b) predicted versus observed values. Note that SSR is calculated as $\text{SSR} = (\mathbf{x}_j - \hat{\mathbf{x}}_j)^2$, where $\mathbf{x}_j$ and $\hat{\mathbf{x}}_j$ are the column vectors containing the sampled and estimated samples of the biomass concentration $j = 6$ of the selected batch.

```
[p_ii_c,t_ii_c] = pca(data_eq_ii_c_as,'NumComponents',2,
    'Centered',0);
```

To observe this effect, we examine the residuals of the first PC of the fitted PCA model on the original data, as shown in Figure 3.9. The residuals distinctly exhibit a dynamic nonlinear behavior throughout the processing. This phenomenon has been frequently reported in the literature.[27, 28, 105, 106] The relationships among the 10 process variables, which must be captured with PCA, vary along the batch processing. Thus, if a single model is fitted, the relationships captured in the model are simply an average, which may not be representative of some parts – or even any part– of the process. A possible solution is to properly split the data into sub-models in such a way that the process behavior can be well approximated with linear techniques and the dynamic relationship structure can be captured.

Example 3.2d: Missing data recovery using a multiphase model

The next example illustrates the implications of estimating intermediate values using the missing data recovery ability of LVMs without considering dynamics but by splitting batch data into multiple phases related to the intrinsic stages of the process. The corresponding code to reproduce this example can be found in the MATLAB live script "3_2d.mlx."

Based on the prior knowledge on the fermentation process, let's start dividing the batch trajectories into the following three phases in the next example: [1, 40], [41, 127], and [128, 183].

```
phase = cell(3,1);
phase{1} = 1:40;
phase{2} = 41:127;
phase{3} = 128:183;
```

Regarding the number of PCs, we select one PC for Phase #1, two PCs for Phase #2 and two PCs for Phase #3.

```
pcs = [1,2,2];
```

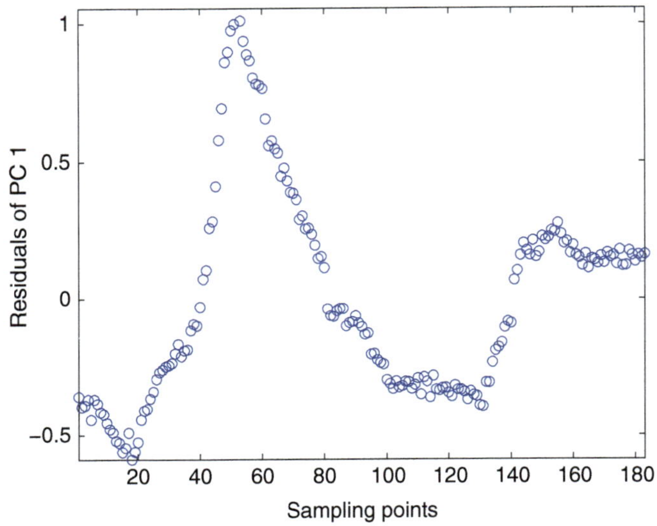

Figure 3.9 Residuals of of the biomass concentration after extracting the first PC of the PCA model fitted on the original data.

The TSR-MB method is applied to each of these phases to impute the missing values.

```
data_eq_ii_d(phase{i},:) = pcambtsr(data(phase{i},:),
    pcs(i));
```

Similarly to the previous examples, we autoscale the raw and the resulting equalized two-way data arrays as follows:

```
data_noneqori_as = preprocess2D(raw_data,2);
data_eq_ii_d_as = preprocess2D(data_eq_ii_d,2);
```

The results of the phase-based estimation of the intermediate values of the biomass concentration are depicted in Figure 3.10a. As observable, this approach leads to better results in terms of prediction (Figure 3.10b), since the associated $SSR = 0.2$ is much lower than in the imputation by TSR-MB ($SSR = 0.468$). These results support the claim that the quality of the estimation of a LVM depends very much on how accurately the time-varying correlation structure is captured.

3.3.2.1 Comparison of Equalization Methods Based on Latent Variable Models

Often, the accuracy of the estimation is not the most important aspect in the analysis or the design of a monitoring scheme. However, when imputing missing values, the aim is to minimize the prediction or imputation error while preserving the covariance structure present in the available data after data equalization. In Figure 3.11, the model obtained from the original data – i.e., full sampling rate of biomass concentration – is compared to the model obtained after the estimation of the biomass concentration by using the different approaches from the previous examples. Slight differences can be appreciated in the resulting latent structures, in particular,

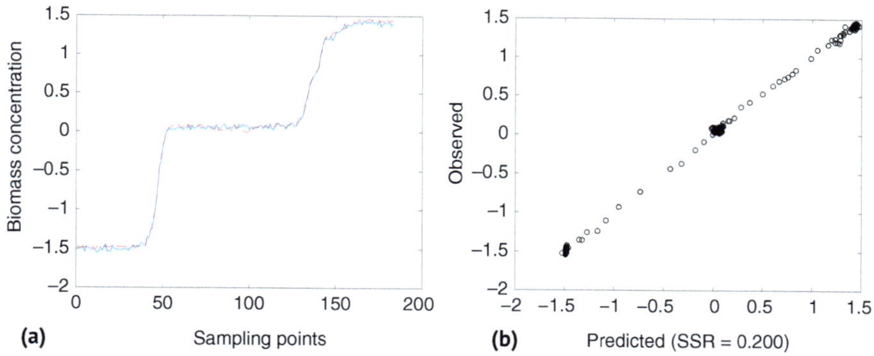

Figure 3.10 Comparison of the original data with data resulting from the estimation of the biomass concentration using the TSR-MB algorithm in a multiphase-based approach, shown in two views: (a) time series and (b) predicted versus observed values. Note that SSR for each of the equalization methods is calculated as $\text{SSR} = (\mathbf{x}_j - \widehat{\mathbf{x}}_j)^2$, where $\mathbf{x}_j$ and $\widehat{\mathbf{x}}_j$ are the column vectors containing the sampled and estimated samples of the biomass concentration $j = 6$ of the selected batch.

between the model obtained on interpolated data (see Figure 3.11a) and the models using the data imputed by using the correlation structure of the process variables (see Figure 3.11b–d). In this example, all approaches show a good approximation in the estimation. The best one is the multiphase approach, but the relative importance of accurately modeling the dynamics (with interpolation or LMVs), the correlation between variables, and the variable nature of the dynamics (with multiphase models) will depend on each case study. Here, we have presented the strategies that can be used for such modeling.

3.3.3 Rearranging Data

The third equalization method proposed involves restructuring the data so that each intermediate measurement is considered as a separate variable, in a similar manner to introducing new lagged variables. This technique enhances the original measurement vectors and, unlike the method consisting in discarding intermediate values, which omits potentially important process information, retains all information, although not in a structured way.

Example 3.3: Rearranging data for equalization purposes

To proceed with the next example, whose code is available in the MATLAB live script "ex3_3.mlx," we start with the same raw batch data used in previous cases, and with the sampling frequency of the biomass concentration modified as described in the previous examples. The next step is to arrange all the process variables except for the biomass concentration in such a way that the intermediate measurements are added as new variables, as shown below:

```
end_v = mod(size(data,1),sampling_frequency);
data_eq_iii = data(1:sampling_frequency:(end-end_v),6);
for i=sampling_frequency:-1:1
```

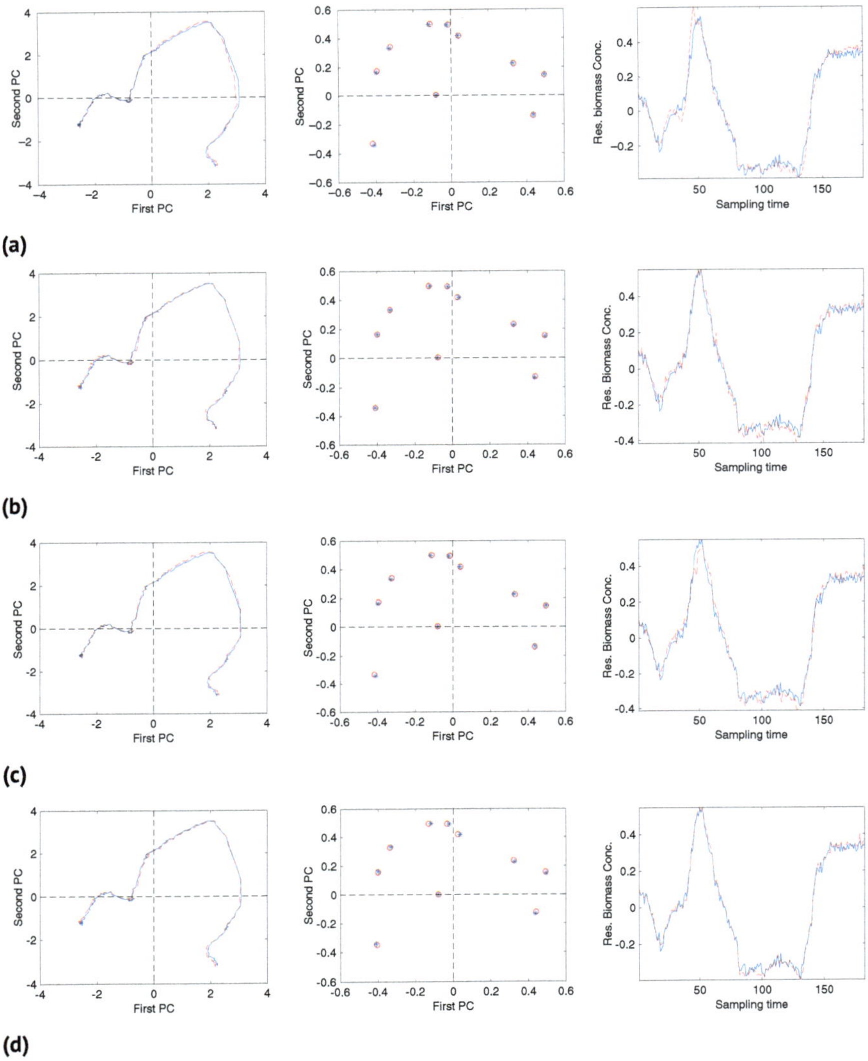

Figure 3.11 Comparison of the PCA models from the original data (blue lines and markers) and the estimated data with different approaches (red lines and markers): (a) cubic spline interpolation, (b) TSR-MB, (c) single-phase dynamic model approach, and (d) multiphase-based approach. First column (scores); second column (loadings); third column (residuals of the biomass concentration).

```
    data_eq_iii = [data_eq_iii data(i:sampling_frequency:
        (end-end_v),[1:5 7:end])];
ends
```

Afterward, we preprocess the raw and the resulting equalized two-way data arrays, respectively, using the autoscaling method:

```
data_noneqori_as = preprocess2D(raw_data,2);
data_eq_iii_as = preprocess2D(data_eq_iii,2);
```

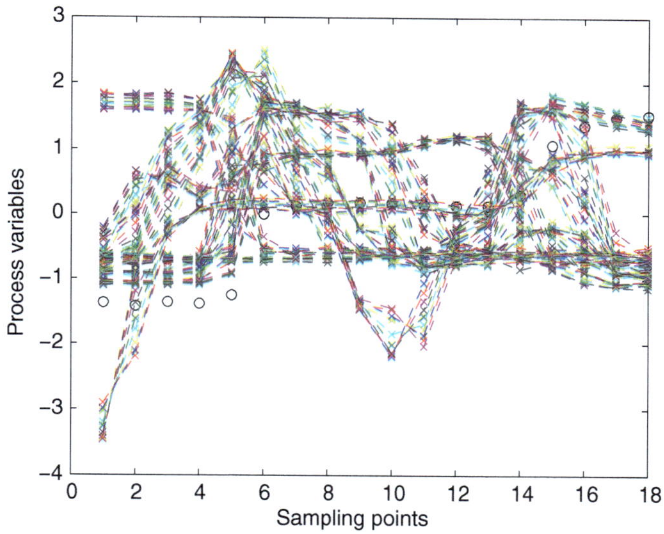

Figure 3.12 Data resulting from the rearrangement of the data.

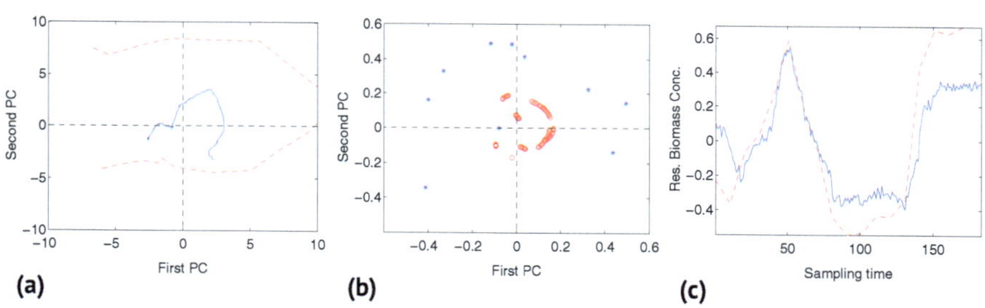

Figure 3.13 Comparison of the PCA models from the original data – no missing values – (blue lines and markers) and the data rearranged (red lines and markers): (a) scores, (b) loadings, and (c) residuals of the biomass concentration.

Finally, we calibrate a PCA model with two PCs on the preprocessed unequalized and equalized datasets, respectively.

```
[p,t] = pca(data_noneqori_as,'NumComponents',2,'Centered'
    ,0);
[p_iii,t_iii] = pca(data_eq_iii_as,'NumComponents',2,
    'Centered',0);
```

The resulting rearranged data are plotted in Figure 3.12. A glance at the figure suggests that this approach may complicate the understanding and interpretation of the model. This notion is reinforced when comparing the model of the original data with the one derived from the rearranged data shown in Figure 3.13. We observe that a completely different model emerges, making interpretation becomes more challenging.

3.4 Multirate System

A multirate sampling scenario, frequently encountered in industrial applications, is typically generated by various factors, including, but not limited to, compression algorithms, variations in raw materials, impurities in feedstock, and other related influences. The primary distinction between such unequalization cases and other sampling scenarios discussed in this chapter is that the number of samples among variables and phases varies across batches. For instance, in the case of nonuniform sampling due to the policy implemented in the compression algorithm, a sample of a specific variable is archived only when a certain time interval has elapsed or when the difference between the current and previous value exceeds a threshold commonly associated with the calibration of the instrument.[175] A good example of this scenario was reported in Ref.[57], where data of the manufacturing of an active pharmaceutical ingredient were used. The objective was to extract valuable information for process understanding and improvement from compressed historical data. In this case study, the median number of samples per variable for the reaction step varied between 65 and 1000 samples, whereas for the distillation step, between 31 and 2001 samples. With the aim of preserving the number of samples and the sampling frequency in each stage, a resampling mechanism was proposed prior to bilinear modeling. This procedure involved resampling each variable based on the ratio of the median number of total samples to the median fraction of total time spent in each stage, using interpolation. The authors claimed that this methodology allowed them to equalize the batch data and also reach some degree of synchronization. However, interpolating this type of unequalized data will not guarantee reconstruction of the time-dependent correlation structure in batch data.[93]

Example 3.4a: Equalization of Multirate systems

The subsequent example illustrates an extreme case of multirate sampling and the performance of interpolation and missing data recovery techniques. The MATLAB code required to reproduce this example can be found in the live script "ex3_4a.mlx."

In particular, the 10 process variables of the *Saccharomyces cerevisiae* cultivation process are sampled with different sampling frequencies among sets of variables: the first five variables were collected every 5 sampling times; variables from 6 to 8 were collected every 3 sampling times, variable #9 was collected every 7 sampling times, and finally, variable #10 was collected every 10 sampling times (see Figure 3.14).

The first attempt to equalize the multirate dataset is to equalize all the process variables by using interpolation based on cubic splines, which will allow us to impute the intermediate values between samples across variables in a univariate fashion.

```
data_eq_ii_a = data;
for v=1:size(data_eq_ii_a,2)
    nnan = isnan(data_eq_ii_a(:,v));
    ind_an = find(1-nnan);
    data_eq_ii_a(:,v) = interp1(ind_an,data_eq_ii_a(ind_an
        ,v),1:size(data_eq_ii_a,1),'spline');
end
```

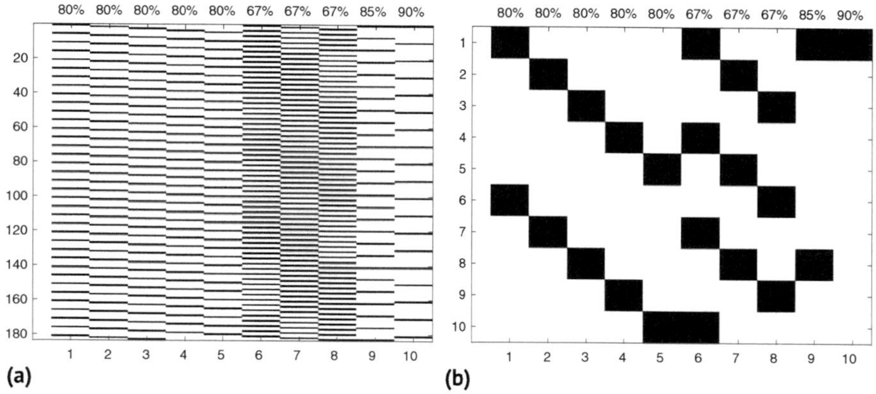

Figure 3.14 Missing data map showing the available (black) and missing (white) measurements for the 10 process variables of the selected batch for the (a) whole batch run and (b) for the first 10 sampling times. The percentage of missing data per variable is shown in the columns of the maps.

For comparison purposes, the multirate dataset is also equalized leveraging the TSR-MB algorithm with three PCs (PCs_opt equal to 3) as follows.

```
data_eq_ii_b = pcambtsr(data,PCs_opt);
```

For visualization and comparison purposes, we autoscale the raw dataset, the raw dataset modified by inducing multirate sampling and the equalized multirate dataset using univariate interpolation and TSR-MB for missing data recovery, respectively.

```
data_noneqori_as = preprocess2D(raw_data,2);
data_noneqmulti_as = preprocess2D(data,2);
data_eq_ii_a_as = preprocess2D(data_eq_ii_a,2);
data_eq_ii_b_as = preprocess2D(data_eq_ii_b,2);
```

As a result, a set of unequalized batch trajectories is generated – see original and resampled data for comparison in Figure 3.15a, b, respectively. The percentage of missing data is excessively high due to the type of multirate scenario simulated, as observed in Figure 3.14. In this situation, when each observation has more missing than available values, the use of the TSR-MB method is not adequate, as illustrated in Figure 3.15c. In order to successfully impute missing values, there must be a certain degree of redundancy in data and collinearity, which is not provided in the dataset of this example. In Ref.[49], a comparative study was conducted among different missing data imputation techniques in scenarios of diverse percentages of missing data without the presence of outliers. One of the conclusions of this study was that the performance of the TSR algorithm considerably degrades when there is a 70–90% of missing data on measurements, assuming the presence of redundancy and high collinearity. In the simulated multirate example, the actual correlation among process variables over time within the batch is not well represented by the updated variance-covariance matrices estimated in each iteration of the TSR-MB algorithm. Hence, the missing measurements cannot be accurately imputed from the measured process variables.

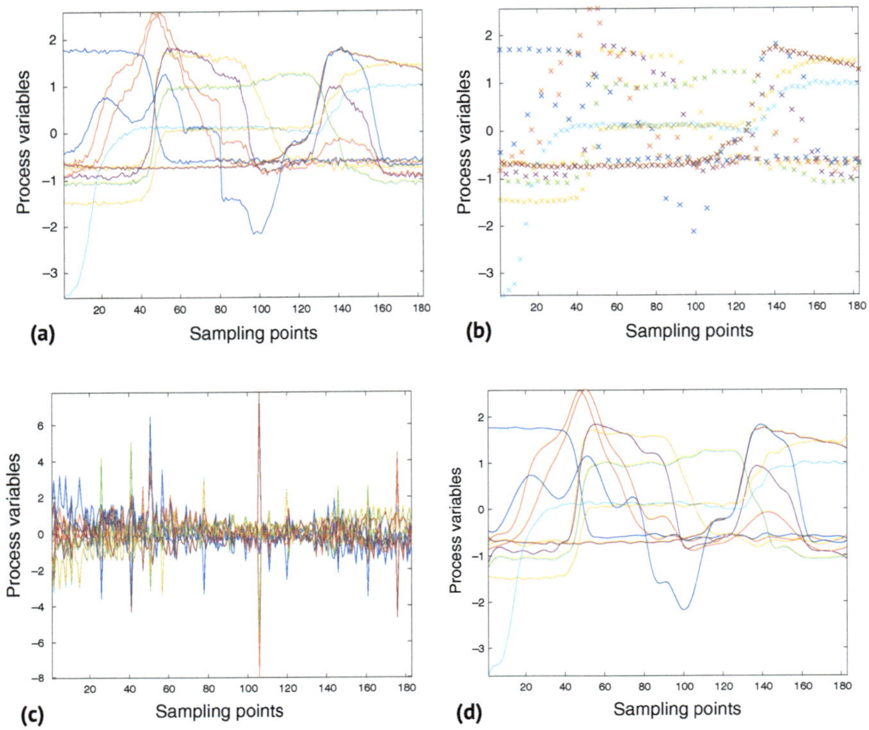

Figure 3.15 Multirate example: (a) original data, (b) multirate data, (c) data recovered by TSR-MB, and (d) interpolated data.

In the simulated multirate scenario, univariate interpolation improves the results obtained from the missing data imputation at a first glance (see Figure 3.15d). However, special caution should be taken with the potential aliasing effect produced by the reconstruction of the trajectory. Comparing the original and resampled data (see Figure 3.15a, d, respectively), there is a loss of information, mainly caused by the sampling frequency at which the data were archived (see Figure 3.15b). Despite the sampling frequency obeys the Nyquist–Shannon sampling theorem, univariate interpolation might add artificial landmarks. Thus, artificial correlation can jeopardize the outcomes of subsequent bilinear multivariate statistical modeling. A clear example of this effect can be observed on the oscillations along the flat profiles of some variables affected by white noise, which are observable in the blue variable trajectories in Figure 3.15d. If there is a risk of distorting the covariance matrix of the data across batches, the corresponding equalization should be discarded.

Example 3.4b: Equalization of multirate systems with consecutive missing values
This example, which is coded in the MATLAB live script "ex3_4b.mlx," addresses an extreme case of multirate sampling so that the number of samples among variables and phases differ accross batches, coupled with the presence of a high number of consecutive samples missing. Notice this scenario is very common in industry since

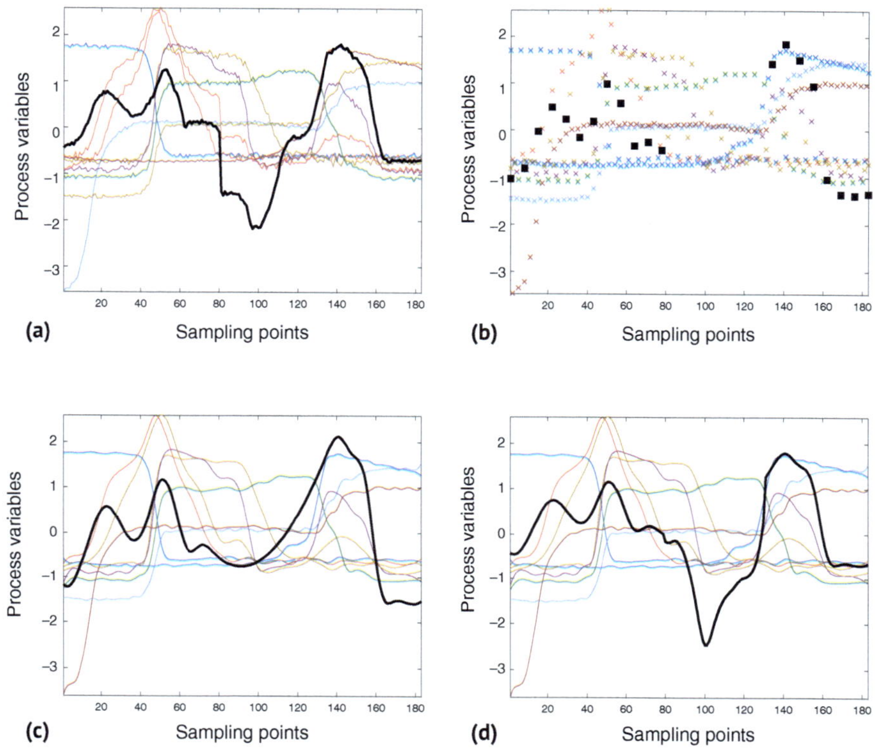

Figure 3.16 Multirate example with consecutive missing values due to a sensor failure: (a) original data, (b) multirate data with missing values in the time interval [80, 130]. (c) interpolated data, and (d) interpolated data with estimation of the missing part by the TSR-MB algorithm.

it is what happens when a sensor breaks down or is malfunctioning. An illustrative example is shown in Figure 3.16.

To start, the multirate dataset generated in the previous example is reused, with the addition of discarding a set of 50 consecutive measurements in the ninth process variable, more precisely, between the 80th and 130th sampling data points (see missing values between the 80th and 130th sampling time points in the trajectory depicted by black squares in Figure 3.16b).

As in the previous example, we start equalizing all the process variables using interpolation based on cubic splines to impute the intermediate values between samples across variables.

```
data_eq_ii_a = data;
for v=1:size(data_eq_ii_a,2)
    nnan = isnan(data_eq_ii_a(:,v));
    ind_an = find(1-nnan);
    data_eq_ii_a(:,v) = interp1(ind_an,data_eq_ii_a(ind_an
        ,v),1:size(data_eq_ii_a,1),'spline');
end
```

To improve the estimation, we proceed with discarding the estimates corresponding to the time period with 50 measurements missing, and subsequently, we impute these missing measurements with the TSR-MB algorithm using three PCs (PCs_opt equal to 3), leveraging the rest of actual and interpolated data.

```
data_eq_ii_b = pcambtsr(data_eq_ii_a,PCs_opt);
```

Similarly to the previous example, auto-scale the raw dataset, the raw dataset manipulated by inducing a multirate sampling, and the equalized multirate dataset using univariate interpolation and TSR-MB for missing data recovery.

Shifting to the analysis of the results, we can observe that the reconstruction using interpolation is not appropriate (see the thick black line in Figure 3.16c). In practice, practitioners do not know whether interpolation is the best method to apply because the original data is not available. In this example, the type of unequalization is known as well as interpolation is prone to error. A way to improve the estimation is to discard the estimates corresponding to the missing period in the ninth variable and try to recover them using the TSR-MB algorithm and the rest of the interpolated data (see thick black line in Figure 3.16d). Although the estimation is not optimum, it outperforms the estimation only considering interpolation. The estimation can be further improved using some of the approaches already discussed in the previous examples, i.e., modeling by phases and incorporating dynamics.

Note that when the sampling rate is different across variables and batches, the unique straightforward solution for data equalization is the interpolation of the samples, as long as the new imposed sampling frequency is compliant with the Nyquist–Shannon sampling theorem, and there is not a significant loss of information that can affect the performance of monitoring systems. The TSR-MB method cannot be applied since there is not a common sampling rate in all batches. Hence, the number of samples might notably vary, and the actual correlation cannot be exploited for missing data imputation. Even though the interpolation method can be applied, special attention must be taken since there might be missing features that this method might not be able to capture. The best practice is to establish a proper sampling rate in the process that takes into consideration sampling constraints (e.g., cost of sampling for subsequent laboratory analyses or importance of process stages), thereby leading to a similar multirate sampling system across batches.

4

Batch Synchronization

4.1 Introduction

The terms "non-equalized" and "unsynchronized" data are commonly confused in the literature.[36, 168] In this manuscript, as explained in Chapter 3, we use the following definitions. The term "non-equalized" is mainly used when the measurements belonging to the process variables are collected at different sampling frequencies within and/or among batches due to manufacturing policies or compression issues. In contrast, process data are unsynchronized when the key process events do not occur at the same time over all batches, irrespective of whether the variables are measured at a constant sampling rate or the batch duration is the same for all batches or not. If key process events are not synchronized, a synchronization procedure must be applied in order to ensure that similar key behaviors overlap in the common time slot across batches. This is referred to in the literature as batch synchronization. In case batch data are non-equalized and unsynchronized, first, an equalization of variables must be performed as illustrated in Chapter 3, followed by the synchronization of the equalized batch trajectories.

In most batch processes, the assumption that all the batch trajectories are synchronized is rarely met. Typically, the recipes for automation are based on triggers that are seldom dependent on time, which cause the batch pace evolution to be different from batch to batch.[57] In addition, different sizes of batch charge, modifications of the recipe to release products at lower costs, impurities in the raw materials, and disturbances in the environmental conditions may produce uneven time-length batches.[187] Hence, not only may the collected batch trajectories for subsequent multivariate data analysis have different lengths, but also the key process events do not overlap at the same time in all batches.[93]

In this context of asynchronous batches, the application of latent variable model (LVMs), such as principal component analysis (PCA) and partial least squares (PLS), is not sensible. One of the strong assumptions of most of these models is that all batch trajectories have equal duration and are synchronized, i.e., similar events happen at the same sampling times.

In the literature, several discussions on the different patterns of time-varying batch trajectories and the possible synchronization solution can be found.[93] In case batches have different durations, but trajectories overlap in the common time part (see Figure 4.1a), a PCA model can be fitted using the information from the long batches while the absent part of the trajectory for the short batches is treated as missing data.[94] For some batches, data collection starts earlier than for other batches, so a shift can be applied to synchronize the batches (see Figure 4.1b). In more complex cases (see Figure 4.1c, d), variable trajectories have different shapes, even when the batch duration is the same, indicating that the timing for key events during each batch is different. Examples of this occur when, for instance, various decisions during the batch are not automated but left to the discretion of an operator, or the process is divided into several stages based on the occurrence of some phenomena that happen at different sampling times. In such cases, more complex techniques are required to carry out the synchronization. By and large, the synchronization of batch trajectories is always advisable, no matter the batch duration, since process events may not be synchronized.

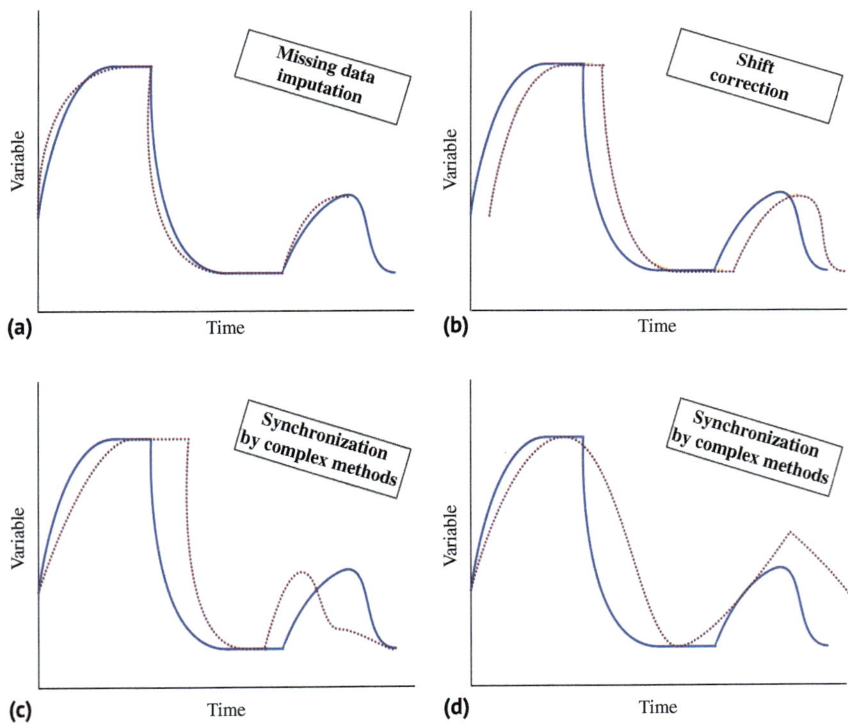

Figure 4.1 Trajectories belonging to one process variable of two batches with different patterns of varying-time trajectories: (a) different batch length with main events occurring at the same time, (b) different starting point yielding a shift in each occurrence, (c) similar batch length with key events do not occur at the same time intervals, and (d) different shape and length.

4.2 Synchronization Approaches

A number of proposals for dealing with the most complex synchronization problems can be found in the literature. As mentioned in Chapter 1, the approaches for batch synchronization can be roughly classified into three categories: (i) methods that compress or expand the raw trajectories using linear interpolation, either in the batch time or maturity index dimension, or an indicator or latent variable dimension; (ii) methods based on feature extraction; and (iii) methods based on stretching, compressing, and translating pieces of trajectories.

Within the first category, some authors have addressed the batch synchronization issue using practical ideas, such as truncating the trajectories of all batches to the shortest batch length, or compressing/expanding the trajectories using linear time adjustments by dividing each sampling time along the trajectory by the time at a certain percentage of the end point.[101, 104, 189] These ideas, although practical, are often inadequate for aligning batch trajectories.[180] Marjanovic et al.[108] proposed to create pseudo-batches based on quality measurements, which were subsequently aligned to the end point. These pseudo-batches were cut to the minimum length, thus generating this way segments of constant length. Thereafter, a revised version of this methodology was proposed, which aligns batch trajectories by identifying short-window LVM at first and then applying these identified models to extend shorter trajectories.[174] However, the last two methods are not appropriate in a context of complex asynchronisms. More sophisticated methods relying on the indirect synchronization of batch trajectories in the latent-variable space were proposed. Eriksson et al.[46] suggested calibrating a PLS model to relate all the process variables at every sampling time point to a dummy variable y, which is created considering the batch length and its relative maturity over the median of the training batch lengths. This maturity index is then used to align the batch trajectories, apart from the subsequent scores and other monitoring statistics, by using linear interpolation, which we call the time linear expanding/compression (TLEC) synchronization method. This approach was implemented in soft independent modelling of class analogy (SIMCA) Release 13.0.3,[165] Umetrics software. Another approach is to characterize each batch trajectory in the latent variable dimension derived from a PCA without exploiting any dynamic, by interpolating to a pseudo-mean score trajectory calculated in each grid element created, which partitions the score trajectories.[177] This approach also results in problems of distorting the trajectories in the synchronization process using interpolation, without considering nonlinear process evolution. Other approaches circumventing the synchronization step for multivariate statistical process control are to use the translation-invariant wavelet decomposition and PCA[132, 133] or modeling the overall historical normal operating conditions (NOC) batch trajectory, which is characterized by individual non-synchronized NOC batches, followed by the development of PCA-based local multivariate statistical process control (MSPC) models that encompass the entire process.[136]

An alternative method used for the alignment of batch trajectories is the indicator variable (IV) approach[118]: "*One way to handle varying batch times in online monitoring is to replace time by another measured variable that progresses monotonically in time and has the same starting and ending value for each batch.*" Monotonicity and a high signal-to-noise ratio are necessary conditions for an indicator variable, but they are not sufficient. To be useful in practice, the indicator variable also needs to represent the mechanism that drives the batch process – this requires process knowledge. The indicator variable can also be a variable computed from some of the measured variables; e.g., extent of reaction.[115] Application of this synchronization approach can be found in several research works.[42, 99, 115, 137, 167, 189, 191] When an indicator variable is not available throughout the batch run, but some process variables can be used as an indicator variable at different process stages, the batch synchronization can be performed stage-by-stage.[55]

If a suitable indicator variable is not available for a given batch process, this type of synchronization cannot be carried out, and other approaches are required. In an attempt to overcome some of the limitations, Sartori et al.[143] proposed a framework based on a tailored indicator variable for each process phase, identified using a surrogate optimization framework to maximize the performance of a predictive model for quality assessment. The eligibility of a correct indicator variable, if it exists, is assessed by: (i) performing a Mann–Kendall test for the monotonicity condition, (ii) guaranteeing a high signal-to-noise ratio by computing an F-test of the variance of the detrended indicator variable versus the non-detrended range, and (iii) checking whether similar initial and end points are found across batches by assessing whether the range of the indicator variable is below the variance of the times corresponding to the initial and end points of the stage. Nevertheless, and despite the partition in batches, the selection of an appropriate indicator variable and the existence of nonlinear paces remain a challenge.

Procedures based on feature extraction were also proposed for batch process synchronization. For instance, a mathematical filter was developed to identify key events in batch trajectories even when these events are not known beforehand.[86] More sophisticated approaches are curve registration[2, 130, 166] and dynamic locus analysis,[153–155] which identify landmarks or special points that characterize process stages and changes (the so-called singular points) in a set of batch trajectories corresponding to process variables, and then, the sample trajectories are warped based on the reference landmarks. Chen and Liu[34] propose to decompose the raw batch trajectories into approximations and details at different scales using wavelets. Contributions from each scale are collected in separate two-way arrays, and data are synchronized at each level using an algorithm based on stretching, expanding, and translating pieces of trajectories. Once synchronized, separate two-way arrays are reconstructed to form new synchronized trajectories.

Warping techniques originally proposed in other areas, such as dynamic time warping (DTW) in speech recognition[141] or correlation optimized warping (COW) in chromatography,[37, 45, 117, 126, 162] have received much attention. A good survey on warping methods for spectroscopic and chromatographic signal alignment can be found in Ref.[12] In recent years, these methods have received much attention in process chemometrics to align and synchronize batch trajectories corresponding to process variables.[52, 61, 62, 88, 103, 128, 152, 195] Kassidas et al. proposed a pseudo-online version of DTW for batch synchronization and also presented guidelines to carry out the real-time synchronization. Nonetheless, this real-time version was shown in a simulation study to be inappropriate for batch monitoring due to the high false alarm rate.[67] The relaxed greedy time warping (RGTW) is a solution to overcome this problem.[67] A derivative dynamic time warping (DDTW) algorithm was proposed to capture the underlying process behavior fingerprinted in the trajectories using derivatives.[90] Nonetheless, noisy data can severely affect the computation of numerical derivatives.[60] Zhang and Edgar[192] proposed a robust DTW algorithm that combines a moving window least squares procedure with derivative DTW to avoid singularity points and reduce the dependency of the results on the reference trajectory. Later on, the robust derivative DTW (RDDTW) algorithm was introduced[193] to cope with singularity point and numerical derivative estimation problems of DTW and DDTW in the presence of noise. Results showed that combining the use of the Savitzky–Golay filter and the DDTW algorithm significantly reduces the number of singularity points and retains the most important features of the original trajectories. Another proposal to deal with the problem of derivatives computation in noisy data is the hybrid derivative dynamic time warping (HDDTW) algorithm,[60] which combines piecewise-linear approximations of the unsynchronized trajectories and DDTW. An important limitation of these methods that is worth highlighting is their inaccuracy in synchronizing in the presence of multiple asynchronisms, which the multisynchro algorithm addresses.[66] Given the larger number of synchronization methods proposed in the literature, we will focus on a subset of approaches, including the most used methods and our suggestions: IV, TLEC, DTW, RGTW, and multisynchro. These methods are reviewed and thoroughly described in the following subsections.

4.2.1 Indicator Variable

The indicator variable synchronization technique is based on the idea of expressing the evolution of a batch as a function of a process variable, or a variable computed from the measurements, instead of the batch time.[119] An indicator variable must fulfill some requirements: it must be strictly monotonic and smooth (not noisy), and have the same starting and end points over all batches, and represent the key process events and driving mechanisms. In case such a variable is available, a synchronization can be carried out by replacing the batch time by the indicator variable. The indicator variable, for instance, can be the amount of monomer or other feeds added to the reactor, substrate concentrations, product concentration, or product yields in bioprocesses. In any case, the choice of the indicator variable depends on the type of process. For a variable to be chosen as an indicator variable, the maturity or percentage of completion of a batch should be represented by such variable. The synchronization using the indicator variable method is performed as follows. Typically, nonuniform increments or indicator-variable levels are selected based on the prior process knowledge to mitigate the risks of univariate interpolation.[94] For instance, these intervals can be defined using polynomial functions of time[191] or the importance of certain process stages in the manufacturing.[115] Nonetheless, equally spaced intervals can also be chosen, provided that the process pace is linear. Once the increments are defined, interpolation is used to transform the batch-time dimension into the indicator variable dimension.

For illustrative purposes, let us qualify the simulation time[1] of the *Saccharomyces cerevisiae* cultivation process as an indicator variable (see Figure 4.2). This variable meets all the indicator variable requirements: smooth, continuous, monotonic, and representative of the events that drive the fermentation process. In regard to the latter requirement, the profile of the indicator variable defines the fermentation evolution. When a value of 19.3 units is reached, the first exponential growth phase ends (see Stage #1 in Figure 4.2). During this phase, glucose is in excess, the microorganism cannot digest the glucose at oxide-reductive growth, and consequently ethanol is formed till maximum concentration. Meanwhile, excretion of pyruvate and acetate occurs till maximum concentration. The transition between the first and second exponential growth phases happens in two different stages. In the first stage, the value of the indicator variable rises until 19.5 units, instant at which the glucose is rapidly consumed and depleted (see Stage #2 in Figure 4.2). In the second stage, pyruvate is completely consumed, whereas acetate is partly used during the oxidative–reductive growth, instant at which the oxygen uptake rate decreases to its minimum and the indicator variable reaches 19.65 units (see Stage #3 in Figure 4.2). At this evolution point, the second exponential growth is triggered, and the ethanol accumulated during the first exponential growth is utilized. Prior to this consumption, ethanol concentration rises till reaching the maximum level in the fermentation, at which the indicator variable shows a value of 22 units (see Stage #4 in

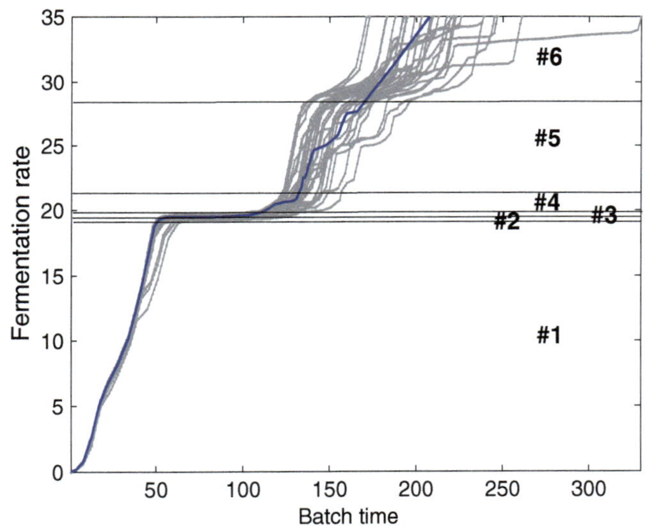

Figure 4.2 Fermentation rate used as an indicator variable for offline batch synchronization of the *Saccharomyces cerevisiae* cultivation process. Nonuniform increments are defined based on the evolution of the fermentation in six different stages of the batch run: #1, 54 increments; #2, 24 increments, #3, 26 increments, #4, 31 increments, #5, 31 increments, and #6: 43 increments. The blue line represents the indicator variable trajectory of a batch whereas gray lines represent the indicator variable trajectories of the remaining batches.

1 Note that the simulation time in the *Saccharomyces cerevisiae* simulator is different (and related in a complex, nonlinear way) to the actual measurement time. The simulation time represents the true evolution of the system.

Figure 4.2). Afterward, the ethanol is consumed, causing the excretion of acetate till the ethanol concentration and the oxygen uptake rate reach a maximum level, where the former is a local maximum. This is the moment at which the second exponential growth finishes, the biomass concentration is maximum in the medium, and the indicator variable rises up to 27 units (see Stage #5 in Figure 4.2). From this point onward, the fermentation enters the stationary phase until the end, when the indicator variable reaches the value of 35 units (see Stage #6 in Figure 4.2).

The synchronization of the fermentation data is performed by selecting nonuniform increments of the indicator variable. In particular, the following indicator-variable levels are defined: 54 equally spaced increments for the indicator variable range at Stage #1, 24 equally spaced increments at Stage #2, 26 equally spaced increments at Stage #3, 31 equally spaced increments at Stage #4, 31 equally spaced increments at Stage #5, and 43 equally spaced increments at Stage #6. Note that for the shortest stages representing the process evolution (Stages #2 and #3), the sampling is considerably higher than in other stages, to comply with the Nyquist–Shannon sampling theorem (see Chapter 3). To complete the synchronization, the process variables are re-expressed as a function of the indicator variable by interpolating the values at the 209 indicator-variable levels (see Figure 4.3).

For online synchronization, the value of the first indicator-variable level or synchronized time point will not be estimated up to a measurement of the indicator value does not exceed the first incremental value. When such point is reached, a linear interpolation is carried out, and subsequently, the data are transformed from the original scale to the indicator variable. The indicator-variable level time is calculated by interpolating the time belonging to the previous and current measurements. This is repeated as long as new measurements of the ongoing batch are available.

Two aspects of the indicator variable approach worth discussing further are the constraints for nominating a process variable as an indicator variable and the resampling procedure to transform the batch-time dimension into the indicator-variable dimension. Some authors assume that imposing a starting and ending value in all batches is a correct procedure to ensure meeting this indicator variable requirement.[36] However, this assumption might severely perturb the outcomes of the synchronization. Other authors assume that defining constant intervals on the indicator variable is sufficient to resample the batch trajectories for synchronization.[55] However, if the indicator variable does not evolve linearly in time, the imposed frequency sampling would not be compliant with the Nyquist–Shannon sampling theorem, the shape of the resulting trajectories would be harmfully affected, and hence, the synchronized data would not be representative of the mechanism driving the process. In order to illustrate the pernicious effects of these policies, a synchronization is performed, taking the biomass concentration of the *Saccharomyces cerevisiae* cultivation process, which is the only variable measured in the fermentation process subjected to be used as an indicator variable (see Figure 4.4). As graphically observable, this process variable does not show a common starting and ending value across batches. To proceed with the synchronization, the batch trajectories are cut to the same ending value in all batches, and constant increments are defined along the indicator variable. The resulting trajectories of this synchronization for glucose concentration, acetate concentration, and

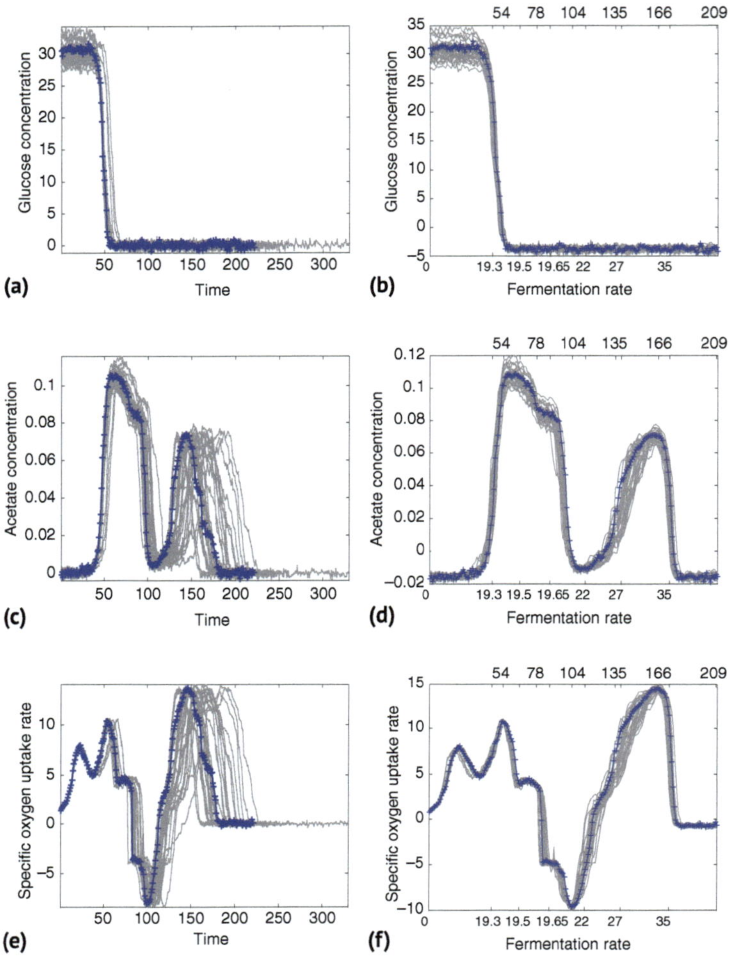

Figure 4.3 Offline synchronization of three variable trajectories belonging to the *Saccharomyces cerevisiae* cultivation process based on the indicator variable-technique: (a, b) glucose concentration, (c, d) acetate concentration, and (e, f) oxygen uptake rate. The raw batch trajectories are plotted as a function of time (a, c, and e) and the synchronized batch trajectories as a function of the increments of the indicator variable – fermentation rate – (lower *x*-axis) and of the number of interpolated samples (upper x-axis) (b, d, and f). Blue crossed lines represent the trajectories of a batch whereas gray lines represent the trajectories of the remaining batches.

oxygen uptake rate are depicted in Figure 4.5. Comparing the raw and synchronized trajectories, apart from observing a downsampling in all the variables, there is a loss of information caused by interpolating at constant increments on the indicator variable. Even though the resulting variable trajectories are equal in length, the loss of the landmarks defining the main process events over the batch run disables the batch data to be used for modeling.

Despite some authors have claimed in the literature that the synchronization based on an indicator variable is the easiest approach to implement and apply to batch data, there are some disadvantages that need to be emphasized. The lack of a process variable that meets the requirements to be an appropriate candidate for a

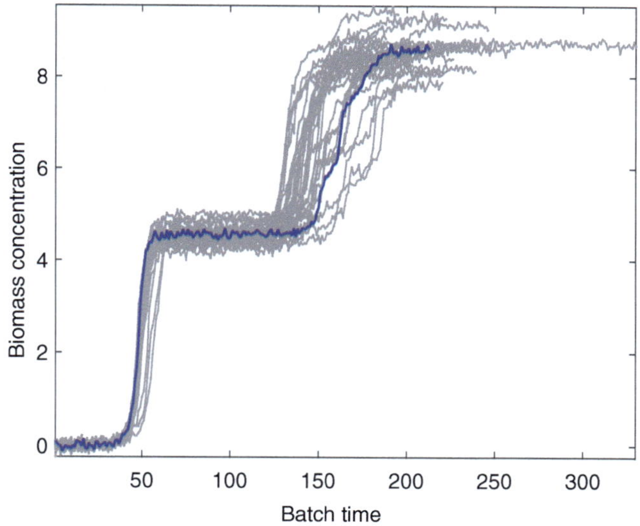

Figure 4.4 Biomass concentration used as an indicator variable for offline batch synchronization of the *Saccharomyces cerevisiae* cultivation process. The blue line represents the indicator variable trajectory of a batch, whereas the gray lines represent the indictor variable trajectories of the remaining batches.

single indicator variable for the entire duration of the batch, or the lack of process knowledge to compute an indicator variable that represents the driving force of the batch process and contains the key process events, are the main pitfalls. Furthermore, this synchronization technique requires expert process knowledge to define the increments to use in interpolation. If the indicator variable levels are not properly determined, interpolation might produce downsampling and compromise the covariance matrix of the data available. Additionally, even though the increments are well designated, interpolation might cause the propagation of outliers and/or the perturbation of the correlation structure.[94] In such cases, more advanced synchronization techniques are required.

4.2.2 Time Linear Expanding/Compressing

The TLEC synchronization[46] is a method integrated in the observation-wise unfolding-T scores batch-wise unfolding (OWU-TBWU) approach[187] and implemented in SIMCA Release 13.0.3[165] for post-batch and real-time batch monitoring. For further details, the different modeling steps of OWU-TBWU are described. Despite the intention of this chapter is not intended to review the use of LVMs for fault detection and diagnosis, a fit-for-purpose explanation on this matter related to the synchronization based on TLEC is provided hereafter. For details on how to model batch data, the reader is referred to Chapters 6 and 7.

The OWU-TBWU approach[187] consists of two levels: the OWU (also called variable-wise unfolding) and the TBWU levels. In the former (OWU level), the main motivation is dimension reduction before the TBWU stage, which is desired in a context of a large number of process variables, e.g., spectroscopic or

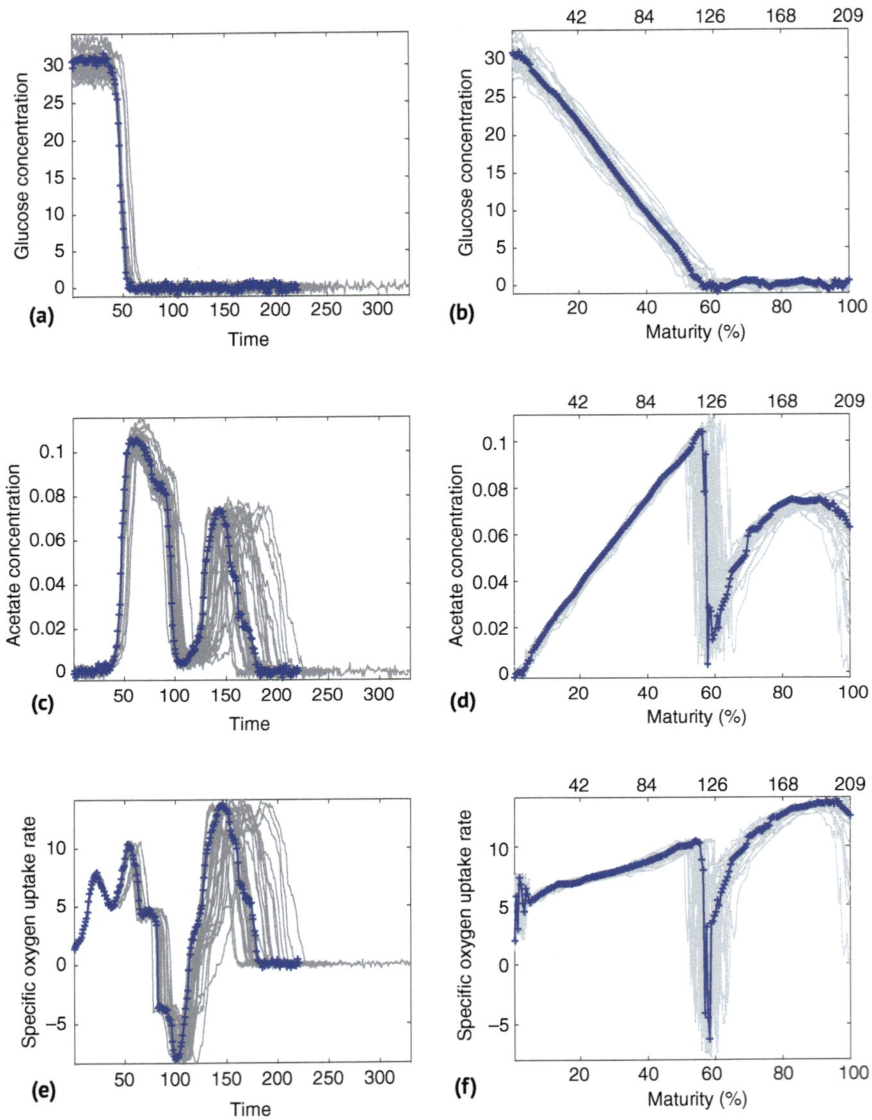

Figure 4.5 Offline synchronization of three variable trajectories belonging to the *Saccharomyces cerevisiae* cultivation process based on the indicator variable-technique: (a, b) glucose concentration, (c, d) acetate concentration, and (e, f) oxygen uptake rate. The raw batch trajectories are plotted as a function of time (a, c, and e) and the synchronized batch trajectories as a function of the maturity of the process based on the biomass fermentation (lower *x*-axis) and as a function of the number of interpolated samples (upper *x*-axis) (b, d, and f). Blue crossed lines represent the trajectories of a batch, whereas the gray lines represent the trajectories of the remaining batches.

chromatographic data. The TBWU level provides a monitoring scheme for end-of-batch online process monitoring based on the OWU scores, along with some multivariate monitoring statistics. Concerning the second level, the aim is to analyze the differences among batches using the information summarized through the matrix scores **T** from the observation level. The observation level (OWU) is described further below, but only the part concerning to the synchronization. For details in this modeling approach, the reader is referred to the original research work.[186, 187]

4.2.2.1 Observation (OWU) Level and TLEC Synchronization Approach

In the OWU level, the three-way array $\underline{\mathbf{X}}$ is unfolded while preserving the variable direction: each variable, measured at different sampling times for the different batches, is arranged in one column of a new two-way array $\mathbf{X}$ $(K_i \times J)$ – K_i being the length of the ith batch – in such a way that each row corresponds to a single sampling time at which the measurements are registered for one batch (see OWU unfolding step in Figure 4.6). Once the arrangement is completed, each column is autoscaled by subtracting its average value and dividing by its standard deviation – the so-called slab-wise preprocessing or variable centering and scaling, as explained in Chapter 5. With this normalization, time periods with less variability will be downweighted, and periods with more variability will be weighted more in the multivariate analysis. Afterward, a PLS model is built in order to relate all the process variables at every sampling time to a dummy variable y, representing the local batch time (see calibration step in Figure 4.6). This variable is created as follows:

1. A vector $\mathbf{y}_i$ containing sorted integer values ranging from 0 to the length of ith batch minus one is built for each single batch.
2. Each value of $\mathbf{y}_i$ is transformed as:

$$\mathbf{y}_{i,k}^{\text{new}} = \frac{\mathbf{y}_{i,k}}{K_i - 1} \times \tilde{K} \tag{4.1}$$

   where $\tilde{K}$ represents the median of the training batch lengths.
3. All the vectors for the different batches are arranged into a single column array, $\mathbf{y}$, containing the values of the variable y, for each batch at each sampling time.
4. $\mathbf{y}$ is autoscaled.

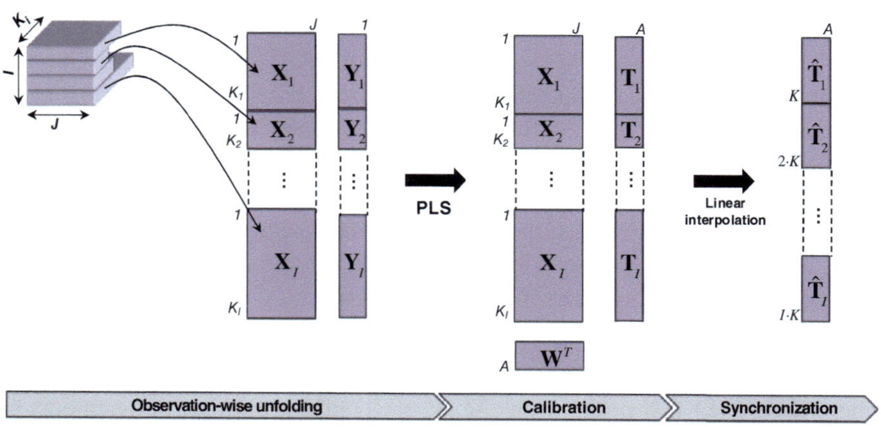

Figure 4.6 Procedure performed in the observation or OWU level. *Source*: Reproduced from Ref.[70], with permission of American Chemical Society.

Note that if all batches have equal duration, the vectors $\mathbf{y}_i$ and $\mathbf{y}_i^{\text{new}}$ contain the same number of elements and the same values. In case the batches have unequal duration, all vectors $\mathbf{y}_i^{\text{new}}$ have the same starting and ending values, but a different number of intermediate values depending on their batch length K_i. Once the PLS model is fitted, the resulting OWU scores $\mathbf{T}_A$, along with some multivariate statistics for each batch, are readjusted by linear interpolation (the so-called TLEC-based method) using the $\mathbf{y}_i^{\text{new}}$ maturity index vector, as outlined in Figure 4.6. This readjustment permits the new OWU scores (denoted as $\widehat{\mathbf{T}}_A$) to be used for process monitoring and subsequent multivariate analysis at the batch level.

A drawback of the observation level in real-time monitoring of new batches is that the OWU scores and the multivariate statistics cannot be aligned until the completion of the batch. Hence, there is no guarantee that the control charts on the OWU stage show aligned results, and therefore, monitoring may be misleading.

Furthermore, it is worth mentioning that the application of the synchronization procedure integrated in the OWU-TBWU approach (the TLEC-based method) may be completely inappropriate due to the underlying assumptions that are rarely fulfilled in batch processes: (i) linear process pace, (ii) all batches are completed and all the key process events defining the process evolution throughout the batch run are present in all batches, and (iii) batches with equal duration are considered as synchronized. If batch data do not meet these assumptions, the process evolution in the trajectories of the process variables will differ batch-to-batch. Hence, the application of methods based on latent structures for process understanding and monitoring is not recommended.

4.2.3 Dynamic Time Warping

DTW is a technique originating in speech recognition, aimed at matching similar events between signals. DTW has the ability to synchronize two trajectories by appropriately translating, stretching, and shrinking sample points to match the reference points in order to minimize the distance between both trajectories.[141]

Kassidas et al.[88] applied the ideas of DTW to batch trajectory synchronization. They adapted the breakthrough work published in Ref.[141] and elucidated the crucial role of constraints in the application of DTW to batch trajectory synchronization. In order to introduce the DTW theory, let $\mathbf{r}$ $(K_r \times 1)$ and $\mathbf{s}$ $(K_s \times 1)$ be the vectors containing the trajectory of one process variable from different batches, measured in K_r and K_s time intervals, respectively. The goal is to make these trajectories equal in length while trying to match the key events. For this purpose, a $(K_s \times K_r)$ grid is firstly built with the trajectory $\mathbf{s}$ in the x-axis and $\mathbf{r}$ in the y-axis (see Figure 4.7b). Let us define the warping function or path as follows:

$$\mathbf{f}^T = \{w(1), w(k), \ldots, w(K_w)\} \ \max(K_r, K_s) \leq K_w \leq K_r + K_s \tag{4.2}$$

where each $w(k)$ is a position on the $(K_s \times K_r)$ grid indicating the elements of trajectories that are matched:

$$w(k) = [\iota(k), \xi(k)],\ k = 1, \ldots, K_w \tag{4.3}$$

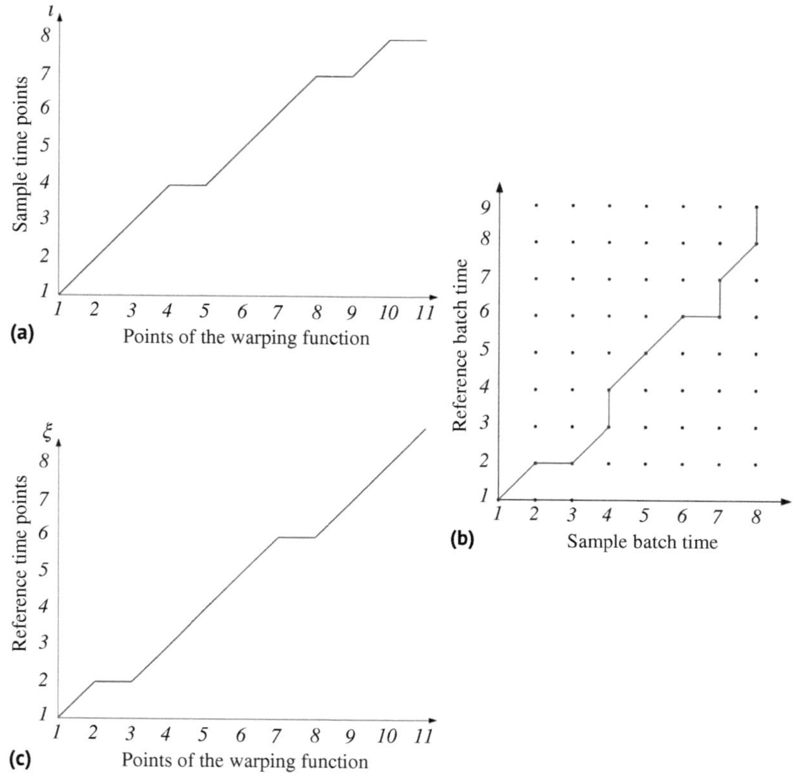

Figure 4.7 Parametric mapping of two different trajectories with different length via the common index *k* (b). Also, a graphical representation of the index row vector $\iota(k)$ (a) and $\xi(k)$ (c) belonging to the sample and the reference trajectories are shown, respectively.

where k is the new index for a common axis that associates the sample with the reference trajectory. Figure 4.7b shows an example how $\iota(k)$ (see Figure 4.7a) and $\xi(k)$ (see Figure 4.7c) are express as function of k. At each grid point, a local distance measured between the $\iota(k)$th sampling time of the **s** vector, $s_{\iota(k)}$, and the $\xi(k)$th point of the **r** vector, $r_{\xi(k)}$, which reflects the dissimilarity between them, is assessed as:

$$d(\iota(k), \xi(k)) = \left(s_{\iota(k)} - r_{\xi(k)}\right)^2 \tag{4.4}$$

Let us define the cumulative weighted local distance between two trajectories along the path **f** as:

$$D(f) = D(\iota(K_w), \xi(K_w)) = \sum_{k=1}^{K_w} d(\iota(k), \xi(k)) \tag{4.5}$$

The aim of the DTW algorithm is to find an optimal path $\mathbf{f}^*$ among all the possible solutions that optimally matches two trajectories, which connect the start and end grid points, in such a way that the cumulative distance D between them is minimized. This is an optimization problem that can be written as:

$$D^*((\iota(K_w), \xi(K_w))) = \min_f [D((\iota(K_w), \xi(K_w))] \tag{4.6}$$

A computationally demanding way to find the optimal path that satisfies (Eq. 4.6) would be to calculate all the possible paths. Once cumulative distances belonging to all possible paths are calculated, the path with the minimum cumulative distance would be the path chosen as optimal. Certainly, this approach would take too much time to obtain the optimal solution. A more efficient approach would be to solve this optimization problem using dynamic programming.[141] This technique is widely used in mathematical optimization and computer programming to solve a complex problems by breaking them down into smaller problems. This method reaches the solution much faster than the approach mentioned above. Before detailing the method to find the optimal path using dynamic programming, certain constraints must be defined to reduce the number of possible paths and the extensive search space.

4.2.3.1 Warping Function Constraints

Several constraints are imposed in the inner DTW algorithm to obtain the optimal path: constraints on the start and end point of the path, monotonicity conditions to keep the temporal order when events occur, local continuity that defines the allowable slope of the path point to point, and global constraints that allow the reduction of the search space.

1. Endpoint Constraints

The end-point constraint imposes the starting $w(1)$ and last point $w(K_w)$ to be (1,1) and (K_r, K_s), respectively. When the start and end points are unknown, this constraint can be relaxed in such a way that an allowable region is defined where the best match of the first and last points on the warping function $\mathbf{f}$ is found. This is particularly useful, especially in online synchronization, where the endpoint at each sampling time is unknown.

2. Monotonicity Conditions

A paramount requirement in the synchronization procedure is to compare events that occur in the natural temporal order, thus reducing the loss of information caused by wrong synchronization. Hence, this constraint can be defined by forcing the path to be monotonous and having nonnegative slopes:

$$\iota(k+1) \geq \iota(k) \tag{4.7}$$

$$\xi(k+1) \geq \xi(k) \tag{4.8}$$

3. Local Continuity Constraints

Furthermore, it is desirable to avoid excessive compressions or expansions of the reference and sample timescale. For this purpose, the local slope of the optimal path is constrained to a range, defining a type of allowable predecessors for each grid point.

Sakoe and Chiba[141] proposed a slope constraint based on the idea of constraining the possible relationship among several consecutive points on the warping function. In case this warping function moves forward to a certain number of m steps in the same direction (x-axis or y-axis), then the following matched points are not allowed to follow further in the same direction before stepping at least n consecutive points in another direction. In order to evaluate the effective intensity of the slope constraint, an intensity measure $In = \frac{n}{m}$ was proposed in Ref.[141] The larger the In is, the more stiffly the warping function will be constrained. If the intensity measure is equal to 0, no constraints are set. If the slope constraint is not sufficiently strict (small value), the discrimination among trajectories will be degraded. In contrast, if In is large, the DTW algorithm will not work effectively. Sakoe and Chiba[141] reported a study with different In and observed that the best result was for a constraint of 1 step both vertical and horizontal, i.e., a slope constraint In equal to one (see Figure 4.8a). Other local continuities used in the literature can be found in Figure 4.8.

4. Global Constraints

One common constraint widely used in the DTW algorithm is the so-called global constraint. Such constraint speeds up the DTW computation and also prevents wrong alignments by globally constraining the area where the optimal path can lie. The most common global constraint regions used are the Sakoe–Chiba band and the Itakura parallelogram, as depicted in Figure 4.9. Note that these global constraints are automatically defined by the local constraints selected. It is due to the imposition of local continuity constraints on the warping function that the region where the optimal path can be located is limited.

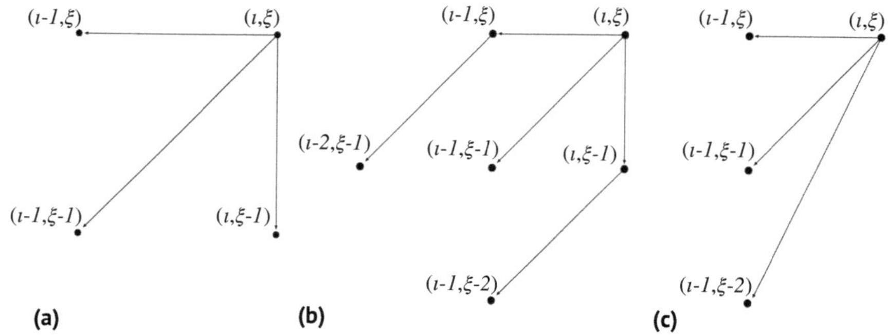

Figure 4.8 Most common local continuity constraints: (a) Sakoe–Chiba local constraint with no constraint on slope, (b) Sakoe–Chiba local constraint with allowable slope in the range $\left[\frac{1}{2}, 2\right]$, and (c) Itakura local constraint allowing slope in the range $\left[\frac{1}{2}, 2\right]$.

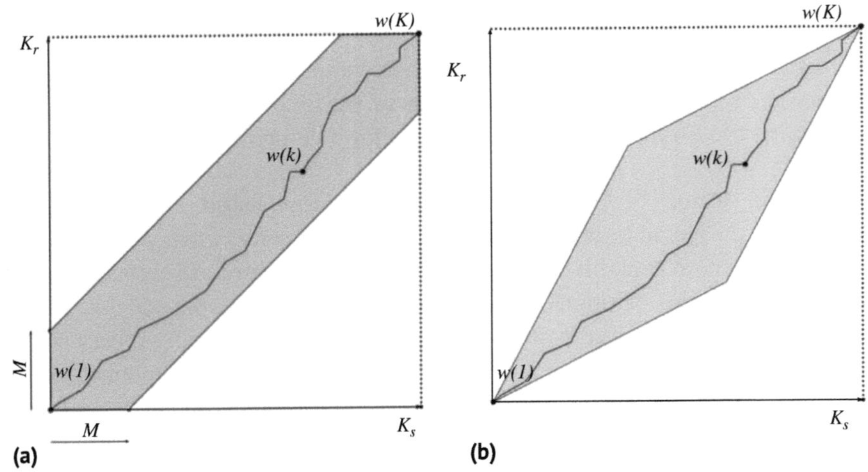

Figure 4.9 Global band constraints: (a) Sakoe–Chiba constraint with width M, and (b) Itakura constraint.

Figure 4.9a illustrates the global constraint band proposed by Sakoe and Chiba,[141] which is defined as:

$$|\iota(k) - \xi(k)| \leq M \tag{4.9}$$

where the parameter M represents the allowable band where the optimal path can go through. This band runs along the main diagonal and constraints the range of warping points to the fixed width of M. In other words, the parameter M defines the length of the vertical vector $\xi(k)$ from the diagonal path to the upper boundary of the band, and likewise, the length of the horizontal vector $\iota(k)$ to the lower boundary. This global constraint is usually used together with the local continuity constraint shown in Figure 4.8a to match trajectories of process variables, avoiding large deviations from the diagonal path.

Regarding the Itakura global constraint illustrated in Figure 4.9b, the search space is defined by a parallelogram, i.e., two lines of slope 1/2 and 2, starting from the start $(1, 1)$ and the endpoint (K_r, K_s), constraining the optimal path search. This is directly implied by the Itakura local constraint, which allows the algorithm to exclude some points on the grid to reach the kth point of the warping function $\mathbf{f}$.

4.2.3.2 The DTW Algorithm

An important aspect to consider when reconstructing the optimal path is the version of the DTW algorithm to use. Namely, there are two versions, the symmetric and asymmetric DTW algorithms. In the symmetric version, the time axis belonging to both the reference and sample trajectory will be transformed into a common axis, where the index k will define the matching point among trajectories. The optimal path will go through all the points in both trajectories, as per the definition in Eq. (4.3). Thus, the outcome of the synchronization will not be different if the reference and sample trajectory are swapped on the axis. Regarding the asymmetric version, it will map the time index belonging to the sample trajectory $\mathbf{s}$ – located on the

x-axis – onto the time index of the reference trajectory $\mathbf{r}$ – located on the y-axis. In this case, the time index k of the signal $\mathbf{s}$ and the optimal path $\mathbf{f}^*$ will contain exactly the same sampling times as the sample $\mathbf{s}$, i.e., K_s. Hence, the optimal path for the asymmetric version is defined as follows:

$$\mathbf{f}^{*T} = \{w(1), w(2), \ldots, w(k), \ldots, w(K_s)\} \tag{4.10}$$

where each $w(k)$ is a position on the $(K_s \times K_r)$ grid indicating the elements of trajectories that are matched:

$$w(k) = [k, \xi(k)] \tag{4.11}$$

This implies that the optimal path will go through all the points of the sample signal $\mathbf{s}$, but it may skip points of the signal $\mathbf{r}$.

There are also symmetric and asymmetric versions of synchronization. In the symmetric version, all the points containing the optimal warping function $\mathbf{f}^*$ are used to synchronize the sample trajectory $\mathbf{s}$. When the optimal path constrains a vertical transition, i.e., from the $(\iota, \xi - 1)$ to the (ι, ξ) point, the response of the signal $\mathbf{s}$ is taken twice. In the case of horizontal transition, i.e., from the (ι, ξ) to the $(\iota - 1, \xi)$ point, the response of signal $\mathbf{r}$ is taken twice. Hence, the length of the synchronized trajectory will have the same length as the optimal path $\mathbf{f}^*$ $(\max(K_r, K_s) \leq K_w \leq K_r + K_s)$. In the case of the asymmetric version, depending on where the reference trajectory is located, either on the x-axis or y-axis, the vertical or horizontal transitions are treated in a different way. For the sake of simplicity, assume that the reference trajectory is located on the y-axis. When consecutive points of the sample trajectory $\mathbf{s}$ are synchronized with the same point of the reference $\mathbf{r}$, which is observable as a vertical transition, the average of these points is estimated and synchronized with the corresponding point of the reference trajectory. In the synchronization of multivariate batch trajectories, Kassidas et al.[88] proposed to use the symmetric DTW followed by an asymmetric synchronization step so that all batch trajectories are synchronized not only with the reference trajectory but also with each other.

4.2.3.3 Optimization Problem

As commented before, the aim of the synchronization of batch trajectories using the DTW algorithm is to find a path that optimally matches a reference with a sample trajectory in such a way that the cumulative distance assessed through all the pairwise points $w(k)$ belonging to the optimal path $\mathbf{f}^*$ is minimized (see Eq. 4.6). Such problem actually becomes an optimization problem because the computational cost associated with assessing all the possible paths to obtain the optimal solution is very high, varying exponentially with the batch length.

To reach this solution in a computationally efficient way, dynamic programming is used to find the optimal solution by searching on the constrained grid under continuity, local, and global constraints, thus reducing computational cost. The procedure to obtain the optimal path can be divided into two phases: the forward and backward phases, as described in Figure 4.10. During the forward phase, the cumulative distance matrix $\mathbf{D}$ is assessed for each (ι, ξ) grid point by solving the

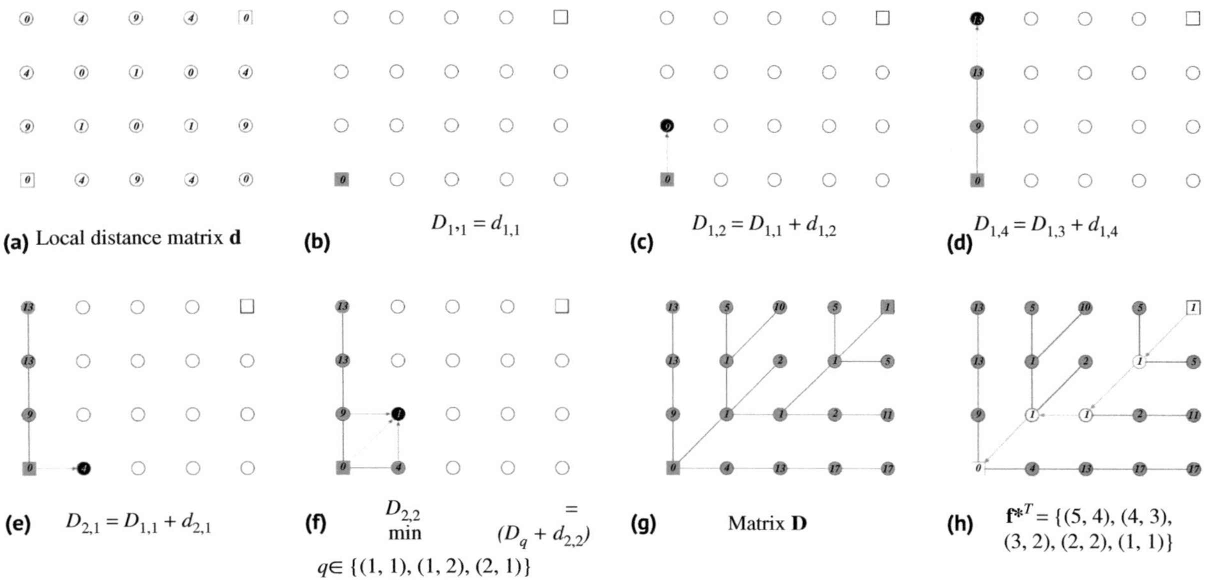

Figure 4.10 Trace of the execution of the symmetric DTW algorithm with Sakoe–Chiba local constraint and no global constraints. Two univariate trajectories, **r** (where $\mathbf{r} = \{1, 3, 4, 3\}$) and **s** (where $\mathbf{s} = \{1, 3, 4, 3, 1\}$), are taken as reference and sample signals, respectively. The dissimilarity among them are expressed by the local distance matrix **d** (see Figure 4.10a). Figure 4.10a–g illustrate the forward phase where the cumulative distance matrix **D** is obtained. Figure 4.10h represents the reconstruction of the optimal path (second phase). A back circle represents the stage that is being evaluated, whereas the gray circles denote the states already evaluated. Dotted arrows show the allowable predecessors for a certain state. The points belonging to the warping function $\mathbf{f}^{*T}$ are depicted by white circles in Figure 4.10h.

following dynamic programming recursive equation:

$$D(\iota, \xi) = d(\iota, \xi) + \min \begin{Bmatrix} D(\iota - 1, \xi) \\ D(\iota - 1, \xi - 1) \\ D(\iota, \xi - 1) \end{Bmatrix} \tag{4.12}$$

Within this first part of the iterative procedure, $D(1, 1)$ is firstly initialized to the value $d(1, 1)$. Secondly, the values of the first column and row of matrix $\mathbf{D}$ are calculated. Afterward, the remaining (ι, ξ) grid points are column-wise estimated depending on the set of predecessors – i.e., $(\iota - 1, \xi)$, $(\iota - 1, \xi - 1)$, and $(\iota, \xi - 1)$ points – and based on Eq. (4.12).

Once the cumulative distance matrix $\mathbf{D}$ is calculated, the optimal path can be reconstructed as a part of the backward phase. The warping function can be found by backtracking from the (K_s, K_r) endpoint to the $(1, 1)$ start point, using the allowable predecessors. Firstly, the predecessor of the endpoint is obtained; subsequently, the predecessor of the previous point is derived, and so forth until the start point $(1, 1)$ is reached. In each step k, the coordinates (ι, ξ) of each predecessor are kept in the warping function $w(k)$, thereby building up the optimal path $\mathbf{f}^{*T}$.

4.2.3.4 End-of-batch DTW Synchronization for Batch Process Monitoring

Let us assume that a set of batch profiles is collected under NOC. Let $\mathbf{X}_i(K_i \times J)$ be the two-way array containing the trajectories of J variables measured at K_i time intervals for the ith batch, where $i = 1, \ldots, I$. DTW works with pairs of patterns; thus, synchronization must be conducted batch-to-batch using a multivariate approach. The aim is to synchronize each $\mathbf{X}_i$ with $\mathbf{X}_{\text{ref}}$. The main steps to carry out in this multivariate synchronization using the DTW algorithm[88] are:

1. **Reference batch selection:** Choose a reference batch $\mathbf{X}_{\text{ref}}$ $(K_{\text{ref}} \times J)$ whose length is the closest to the median duration of the collected NOC batches. The median is less sensitive to outliers, and thus it is preferred to other choices such as the average.
2. **Scaling:** Scale the batches in order to normalize the differences in the units used to record the trajectories of the J variables. Calculate, for each variable trajectory, its average range $\bar{R}_j$ by averaging the range from each batch as follows:

$$\bar{R}_j = \frac{\sum_{i=1}^{I} \max(\mathbf{x}_{i,j}^T) - \min(\mathbf{x}_{i,j}^T)}{I} \tag{4.13}$$

where $\mathbf{x}_{i,j}^T(1 \times K_i)$ is the row vector of the two-way array $\mathbf{X}_i$.

Then, each variable in all batches is divided by its average range as follows:

$$\widetilde{\mathbf{x}}_{i,j}^T = \frac{\mathbf{x}_{i,j}^T}{\bar{R}_j} \tag{4.14}$$

If batch data have the same measurement units across variables, as in spectroscopy data[129] or in data collected from artificial tongues and noses, the scaling step for DTW can be omitted.

3. **Applying the symmetric DTW algorithm with warping function constraints:**
   i. Build a $(K_i \times K_{\text{ref}})$ grid for all the sampling times of $\mathbf{X}_i$ and $\mathbf{X}_{\text{ref}}$.
   ii. Estimate the weighted local distance between each time point $\iota(k)$ of the two-way array $\mathbf{X}_i - \widetilde{\mathbf{x}}^T_{i,\iota(k)}$ $(1 \times J)$ – and each time point $\xi(k)$ of the two-way array $\mathbf{X}_{\text{ref}} - \widetilde{\mathbf{x}}^T_{\text{ref},\xi(k)}$ $(1 \times J)$ – that are allowed by the upper u and lower l boundaries. Such distance is estimated as follows:

$$d(\iota(k), \xi(k)) = \left(\widetilde{\mathbf{x}}_{i,\iota(k)} - \widetilde{\mathbf{x}}_{\text{ref},\xi(k)}\right)^T \cdot \mathbf{W} \cdot \left(\widetilde{\mathbf{x}}_{i,\iota(k)} - \widetilde{\mathbf{x}}_{\text{ref},\xi(k)}\right) \tag{4.15}$$

   where $\mathbf{W}$ is a $(J \times J)$ diagonal nonnegative matrix used to give more weight to certain variables.
   iii. Assess the cumulative distance matrix D applying the forward DTW algorithm.
   iv. Obtain the optimal path $\mathbf{f}^*$ by backtracing from the (K_i, K_{ref}) endpoint up to the $(1, 1)$ starting point, taking the optimal predecessor in each intermediate point.
4. **Apply the asymmetric synchronization:** The points of $\mathbf{X}_i$ that are synchronized with the same point of $\mathbf{X}_{\text{ref}}$ shall be replaced with the mean value of the original values of $\mathbf{X}_i$.
5. **Repeat Steps 3 and 4 for each batch.**

As previously mentioned, the weight matrix $\mathbf{W}$ is used to give more importance to certain variables based on a particular criterion. When process knowledge is available, the weights are set to certain values. However, this is not typically common in industry. As an alternative, the synchronization of batch trajectories should rely more on variables that are more consistent from batch to batch and less on noisy variables.[87] A procedure was proposed to automatically detect, and increase the weight of consistent variables and decrease the weight of the rest in an iterative manner until convergence.[88] This procedure is carried out as follows. First, the DTW algorithm is applied to all batches at each iteration. Afterward, the sum of the squared deviations from the average trajectory of each variable over all synchronized batches is calculated as an index of its consistency. The inverse of this index will be the weight given to a particular variable. The synchronization of the batches is repeated until the weights of the variables converge. One drawback of this weighting criterion is that consistent horizontal trajectories, with no warping information, might be given more weight to seek the optimal path. Another shortcoming is that process variables with a more dynamic behavior shown as time-varying smooth curves – which are less consistent among batches but important for synchronization – might be given less weight. This fact might affect the quality of synchronization even though a phase-to-phase synchronization is completed, leading to inaccurate alignments of the events that drive the batch process.

In the field of analytical chemistry, elution time variation is a common phenomenon in chromatographic data, and this must be tackled prior to modeling. Spectra are curved-like variables that can be divided into three constituting parts: the analytical relevant signal in the form of peaks of different sizes and shapes, the background or baseline, and the noise. Due to subtle, random, and often unavoidable variations in instruments, these signals are usually time-shifted as well. To align the smooth peaks in time, a variant of the DTW algorithm was proposed by Ramaker et al.,[128] which focuses on aligning the spectra features, giving more weight to those variables or wavelengths containing more warping information. This method consists of estimating which variables are more important from a warping perspective in an automatic manner. For this purpose, a weight matrix is initialized to the identity matrix. The first batch is selected and synchronized with the reference batch. Local distance matrix for each process variable between the first batch and the reference batch is obtained. Once all local distance matrices are available, a surface plot is depicted for each matrix, and the optimal path for each variable is mapped on the surface plot. Important warping information is expected to fall in valleys, as illustrated with a white line in Figure 4.11. In other words, the location of the optimal path should be near a valley and close to the diagonal, characterized by small distances. To quantify the importance a variable should have in the synchronization procedure, the ratio of the mean local distance of the elements not lying on the optimal path to the mean distance of the coordinates of the optimal path was proposed. This weight provides a measure to quantify whether a variable is important or not. If this index is high, it means that the path follows the valley of the surface plot, and hence, the mean local distance of the points lying on the optimal path shall be smaller compared to those outside the path. This whole procedure is repeated for each batch to calculate the weights of the J variables. By averaging the weights for each one of the J variables across batches, the diagonal elements of the weight matrix $\mathbf{W}$ are obtained. The complete procedure to calculate $\mathbf{W}$ is repeated until convergence.

In summary, there are two main approaches to select the weights of $\mathbf{W}$ for DTW synchronization. The first approach, proposed by Kassidas et al., gives more importance to variables that are more consistent from batch to batch and less to noisy variables. The second approach, published by Ramaker et al., gives more weight to those

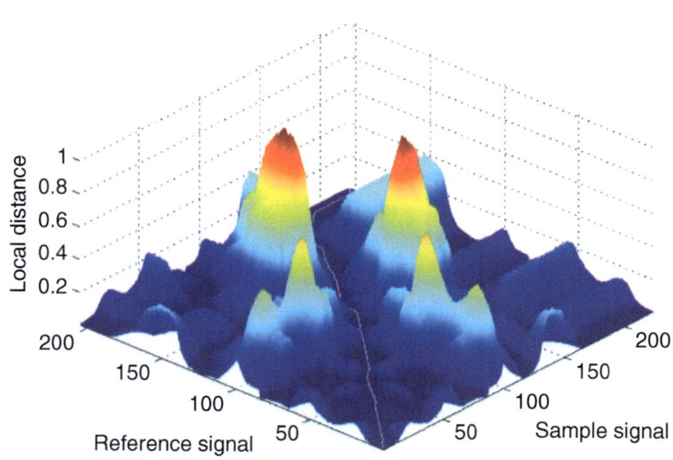

Figure 4.11 Optimal path estimated without global constraints and mapped on the surface plot of local distances for a process variable.

variables containing more warping information. Ideally, the variables that fulfill the requirements of these two methods should be weighted more to accurately synchronize, which suggests a new weight definition:

$$w_{j,j} = \sqrt{w_{j,j|\mathrm{Ram}} \times w_{j,j|\mathrm{Kass}}} \tag{4.16}$$

where and $w_{j,j|\mathrm{Kass}}$ are the geometric means of the weights estimated following the Ramaker et al.'s approach and Kassidas et al.'s approach for each variable, respectively. The geometric average is used because it gives a good measure of the central tendency of the set weights that are calculated based on different criteria and might have different ranges.

Note that in all the above approaches, $\mathbf{W}$ is normalized in such a way that the sum of the diagonal elements is equal to the number of process variables J, to avoid giving more importance due to differences in magnitude of the resulting calculated weights.

4.2.3.5 On the Use of Warping Information

The warping profiles obtained from synchronization are composed of a set of three different transitions at each sampling time, i.e., vertical, horizontal, and diagonal transitions. Based on the number of different transitions the warping path contains, conclusions regarding the performance of the different process stages can be drawn. Let us assume that the sample and reference batches are located on the x-axis and y-axis, respectively. An excessive number of vertical transitions in the warping profile means that the sample batch needed less time than the reference batch for completion. In contrast, an excessive number of horizontal transitions is related to a slow process pace of the sample batch in comparison to the reference batch. To correct these differences in the process pace, the DTW algorithm expands and compresses the pieces of trajectories in such a way that the key process events are synchronized across batches. Note that the warping profiles contain valuable information about the duration of the process substages, which may be associated with anomalies occurring in the process and/or the quality of the final product. Hence, the use of the warping information for process monitoring is highly recommended.[55, 67, 71]

Example 4.1: Synchronization based on DTW

This example illustrates the execution of the DTW algorithm for end-of-batch process synchronization using the three methods described in the previous section, namely: the ones by Kassidas et al., Ramaker et al., and the combination of the two methods. The complete sequence of MATLAB® instructions to reproduce the example and graphical results can be found in the MATLAB live script "ex4_1.mlx."

The first step consists of selecting a batch from the calibration dataset to align the remaining batches. If prior knowledge about the process is available, we can choose a specific batch as a reference. The function ||`refBatch`|| of the MVBatch toolbox can select the batch whose length is closest to the median or the average duration of the collected batches. In this particular case, the median will be selected, thereby guaranteeing a representative batch of the typical batch-to-batch variation is used.

```
reference_batch = refBatch(raw_data,'median');
```

An aspect to be taken into account is which variables should be introduced into the DTW algorithm. It may happen that there are on/off variables providing information about different episodes or phases in the process that would be extremely useful in the synchronization step to improve the synchronization. In this example, all the measured variables are used for synchronization. A binary array can be defined to pass it on to the synchronization method. This array will have as many integers as the number of process variables; 1s imply that the corresponding variable is constrained, and therefore, it will not be used in the computation. Note, however, this step is not necessary if all the process variables are considered in the computation of the weights for synchronization, because the algorithm will define this array with all 0 values by default. However, if the practitioner would like to discard one variable, the corresponding position in the array would have to be set to 1. In this example, all the process variables are considered; therefore, the definition of the variable constraints can be stated as follows:

```
variable_constraints = zeros(size(raw_data{1},2),1);
```

In the next step, we will proceed with the synchronization of the batch trajectories using the iterative procedure proposed by Kassidas et al. More precisely, the algorithm automatically detects and increases the weight of consistent variables, and decreases the weight of the rest in an iterative manner until convergence. For the execution of this sequence of steps, we can proceed with the synchronization:

```
[weights_kass, synchronized_data_kass,
    warping_profiles_reference_kass, ~,
    warping_profiles_raw_kass] = ...
     DTW_Kassidas(raw_data,raw_data{reference_batch},
        variable_constraints);
```

A second alternative is to calculate the weights of the synchronization, considering that those variables that can provide more warping information should be candidates to drive the alignment of the batch trajectories. Despite the approach proposed by Ramaker et al. being conceived for the alignment of spectra data, there is merit in compressing and stretching the original trajectories, taking as a reference, variables containing more "profile information" related to key process events.

```
[weights_ram, synchronized_data_ram,
    warping_profiles_reference_kass, ~,
    warping_profiles_raw_kass] = ...
     DTW_Ramaker(raw_data, raw_data{reference_batch},
        variable_constraints);
```

The last approach to estimate the weights of the synchronization is to consider that the best weighting of the variables is the one that takes into consideration both the relevant batch-to-batch variation representative of the NOC in the process and those variables which are more relevant to accurately align batch trajectories. For this purpose, the geometric mean of the weights derived from the DTW synchronization following Kassidas et al.'s and Ramaker et al.'s approaches is calculated as follows:

```
weights_geom = sqrt(weights_ram.*weights_kass)/sum(sqrt(
    weights_ram.*weights_kass))*size(raw_data{1},2);
```

As explained in Section 4.2.3.4, Kassidas et al.'s approach gives more weight to those variables more consistent from batch to batch, i.e., those variables whose trajectories are not notably deviating from the average trajectory. These variables are characterized by a high signal-to-noise ratio, therefore, high weights are assigned to them. In this example, the iterative procedure gives less weight to glucose, acetaldehyde, acetate, ethanol, and biomass concentration, where the last two variables are the least weighted (see the lowest weights among those marked as magenta asterisks in Figure 4.12). However, the most weighted variables are the specific oxygen uptake rate and carbon dioxide evolution rate (see the highest weights among those marked as magenta asterisks in Figure 4.12). This weighting is also illustrated in Figure 4.13, where the least (biomass concentration), intermediately (active cell material and acetaldehyde dehydrogenase), and most weighted variables (specific carbon dioxide evolution rate) are plotted after offline DTW synchronization. The biomass concentration – the 6th process variable – depicted in Figure 4.13a, has a low signal-to-noise ratio in certain stages. This weighting outcome is caused by the high variation observed throughout the entire batch, leading to a low weight in the synchronization. On the contrary, the specific carbon dioxide evolution rate – the 10th process variable, plotted in Figure 4.13d shows little variation around the average trajectory, so a high weight is assigned. The active cell material and acetaldehyde dehydrogenase variables, however, fall into an intermediate weighting category, where larger differences are appreciated as per the weights assigned to the 7th and 8th process variables, shown in Figure 4.13b, c. In the case of the active cell material, there is not a notable variation between the average trajectory and the training batches from the

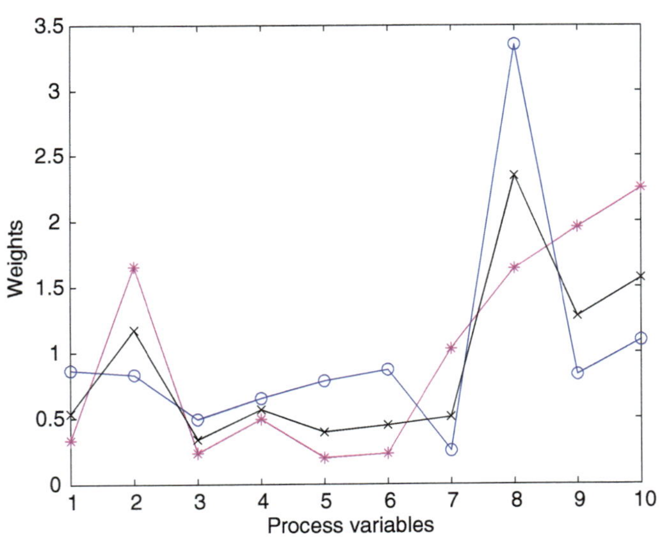

Figure 4.12 Values of the weight matrix **W** assessed according to the approaches of Kassidas et al.'s (magenta asterisks) and Ramaker et al.'s (blue circles), as well as the geometric mean of both weigths (black crosses).

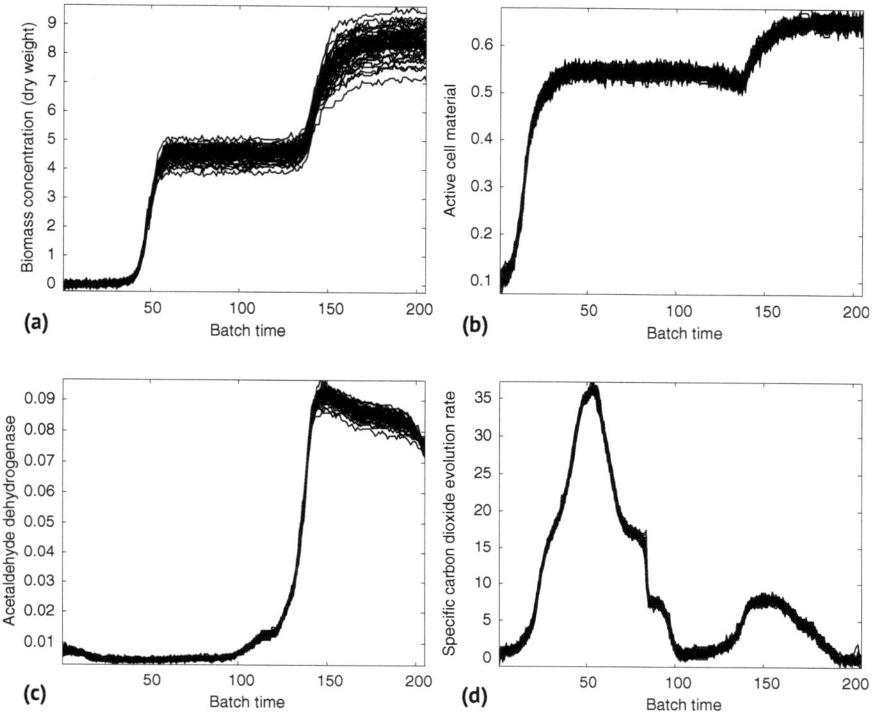

Figure 4.13 Trajectories belonging to four process variables after offline synchronization using DTW based on the weights calculated as the geometric mean of Ramaker et al.'s and Kassidas et al.'s weights: (a) biomass concentration (dry weight), (b) active cell material, (c) acetaldehyde dehydrogenase, and (d) specific carbon dioxide evolution rate.

start to the 40th sampling time, but from the 50th sampling time onward, the variability around the mean increases, which implies a reduction of the signal-to-noise ratio. Regarding the 8th variable, acetaldehyde dehydrogenase seems to be consistent from the beginning to the 150th sampling time, but after this sampling time, a notable variation is observed. Considering these observations, these last two variables are given an intermediate weight, being lower on the 7th variable than on the 8th variable, as evidenced by Figure 4.12.

According to Ramaker et al.'s approach, the weights should be given in relation to the amount of warping information contained in the variables. The weights calculated by following this approach are depicted by blue circles in Figure 4.12. This plot shows that variables with less warping information are the acetaldehyde concentration and the active cell material variables (the 3rd and 7th process variables, respectively), whereas the variable acetaldehyde dehydrogenase (the 8th process variable) has a high warping information content. Furthermore, it is worth emphasizing the fact that the 8th process variable receives more weight following the approach of Ramaker et al. than the approach of Kassidas et al. The cause of this difference lies in the shape of the 8th variable trajectory, which contains the most important occurrences in the process. Hence, this variable serves as a benchmark to achieve a more accurate synchronization of the rest of the process variables.

In order to assign weights more appropriately, the weights from both approaches are geometrically averaged. The purpose is to give more weight to those variables that are more consistent from batch to batch and that also incorporate valuable warping information. As observable in Figure 4.12, those variables that contain more warping information but are not consistent from batch to batch or vice versa, are weighted to an intermediate value between Ramaker et al.'s and Kassidas et al.'s weights. In case variables are assigned similar weights following the approach of Ramaker et al. and Kassidas et al., it leads to similar weights by the geometric mean approach (see Eq. 4.16).

It is worth emphasizing that the weights of the synchronization may vary and differ considerably among datasets of the process because the variability in the batch trajectories will downweight or upweight accordingly based on the variability observed.

After calculating the weights, the next step is to synchronize the batch trajectories using the DTW algorithm.

```
[raw_data_scaled,rng]= scale_(raw_data);

[synchronized_data_geom,
warping_profiles_reference_geom,
warping_profiles_raw_geom]= ...
DTW(raw_data_scaled,raw_data_scaled{reference_batch},
diag(weights_geom),0,
variable_constraints);
```

4.2.3.5.1 Real-time DTW Synchronization for Batch Monitoring

Kassidas et al.[88] proposed an approach to carry out the real-time synchronization for online batch process monitoring. Let $\mathbf{X}_{ref}$ be the two-way array of the reference batch and $\mathbf{W}$ the weight matrix used to carry out the offline synchronization of the NOC batch data. When new measurements are available at the tth sampling time, each variable is scaled to the average range estimated from the trajectories of NOC unsynchronized batches. Let $\mathbf{X}_{new}$ $(t \times J)$ be the trajectories of J variables measured from the start of the batch until the t current time of an ongoing batch.

The use of the offline DTW algorithm for online synchronization is inadequate because the endpoint is unknown at each sampling time. Therefore, the endpoint constraint must be relaxed in order to obtain the optimal path that provides the best matching between $\mathbf{X}_{ref}$ and the ongoing batch $\mathbf{X}_{new}$.

Following the above idea, a set of cumulative weighted distances $D_{t,j}$ at the tth sampling time is calculated, where $j = l_t, ..., u_t$ are the allowed points of $\mathbf{X}_{ref}$ imposed by the lower l_t and upper u_t bound at the tth sampling time. Let us define e_t^* as the sampling time of the batch reference $\mathbf{X}_{ref}$ where the minimum cumulative weighted distance $D_{t,j}$ occurs at the tth sampling time. Hence, the optimal path with the minimal total distance up to the e_t^* point can be determined. After applying the previous procedure, an asymmetric synchronization is required to have a length of e_t^* sampling times. Once the ongoing batch is synchronized, it can be monitored online using different approaches.[187]

One important drawback of this real-time synchronization procedure is the search of the optimal path at each sampling time. Performing a backward search when a new sample is available in order to find the best matching path from the current sampling time t to the start time (1,1) may derive a different optimal path from the synchronization carried out from the previous sampling time t-1 to the start point. This allows the procedure to search for the optimal global path as the batch evolves, but since this is done in real-time, the batch trajectories synchronized at the later stages may differ from those obtained from the synchronization at the earlier stages. The projection of these trajectories onto the latent space will lead to different values in the multivariate statistics that may cause uncertainty and false alarms in the monitoring scheme.[63] Another challenging aspect is the adaptation of the global constraints in real-time synchronization. Since the duration of incoming batches is unknown and may differ from those batches used to tune up the parameters of the algorithm, the update of the upper and lower boundaries is desirable. If this global constraint is not updated on demand, erroneous synchronizations may be produced with the subsequent increase of the false alarm rate.[63] González et al.[68] proposed the RGTW algorithm for real-time batch synchronization that attempts to solve the limitations of this synchronization procedure. This algorithm is further described in the next section.

4.2.4 Relaxed Greedy Time Warping

The RGTW algorithm builds up a piecewise solution using a greedy optimization approach, where the best local synchronization improvement is iteratively integrated into the global synchronization solution. This method is inspired by the work of Kassidas et al.[88] However, unlike their approach, the RGTW algorithm synchronizes within a cross-validated sliding window ζ of defined width γ (time points). The search space is constrained by limits calculated based on the variability of synchronization observed in NOC batches at each time, which are time-varying and adaptable to synchronization variability of new batches.

The execution of the RGTW algorithm, as outlined in the pseudo-code presented in Algorithm 1, requires several input parameters: (i) a vector of measurements of an evolving batch $\mathbf{x}_{\text{new},t}^T$ $(1 \times J)$ at the tth sampling time, (ii) the global constraints defined by the upper $\mathbf{u}^T$ $(1 \times \max(K_{w_i}))$ and lower $\mathbf{l}^T$ $(1 \times \max(K_{w_i}))$ limits (see Section 4.2.4.1 for definition of global constraints customized for batch synchronization), (iii) a window ζ with a certain width γ, which can be defined by cross-validation (see Section 4.2.4.2 for details in the optimization of the window width by cross-validation), (iv) weight matrix $\mathbf{W}$ $(J \times J)$ and (v) the cumulative weighted distance matrix $\mathbf{D}$ $(t - 1 \times K_{\text{ref}})$. As a result of the RGTW synchronization, the algorithm returns the synchronized batch data $\widetilde{\mathbf{x}}_{\text{new},t}^T$ $(1 \times J)$ along with the corresponding optimal path $\widehat{\mathbf{f}}_t^T$.

Figure 4.14 illustrates the main idea of the online synchronization of an ongoing batch, assuming a window ζ with a width $\gamma = 3$, with boundaries $\mathbf{u}$ and $\mathbf{l}$ defined from a previous synchronization using DTW. In this example, when a new vector of measurements is available at each sampling time, the end point, hereafter defined as e^* – i.e., the grid point in the window where the minimum cumulative weighted

Algorithm 1 RGTW Algorithm

Input: vector of batch data $\mathbf{x}_{\text{new}}^T (1 \times J)$ at the tth sampling time, upper $\mathbf{u}^T$ $(1 \times \max(K_{w_i}))$ and lower $\mathbf{l}^T$ $(1 \times \max(K_{w_i}))$ limits, window width γ, weight matrix $\mathbf{W}\,(J \times J)$ and cumulative weighted distance matrix $\mathbf{D}\,(t - 1 \times K_{\text{ref}})$.

Output: Synchronized data $\widetilde{\mathbf{x}}_{\text{new},t}^T$ $(1 \times J)$ and final optimal path $\widehat{\mathbf{f}}^T$ $(2 \times K_{w_{\text{new}}})$

Step 1: Scale batch data

$\widetilde{\mathbf{x}}_{\text{new},j}^T = \dfrac{\mathbf{x}_{\text{new},j}^T}{\bar{R}_j}$, where $j = 1, \ldots, J$ and $\bar{R}_j$ is estimated as per Eq. (4.13).

Step 2: Apply the symmetric DTW algorithm with constraints

Estimate the weighted local distances $d_{t,\xi(k)}$, where $\xi(k) = \hat{l}_t, \ldots, \hat{u}_t$

$d_{t,\xi(k)} = \left(\widetilde{\mathbf{x}}_{\text{new}} - \widetilde{\mathbf{x}}_{\text{ref},\xi(k)}\right)^T \cdot \mathbf{W} \cdot \left(\widetilde{\mathbf{x}}_{\text{new}} - \widetilde{\mathbf{x}}_{\text{ref},\xi(k)}\right)$

Compute the cumulative weighted distances $D(t, \xi(k))$

$D(t, \xi(k)) = d(t, \xi(k)) + \min\{D(t-1, \xi(k)),\ D(t-1, \xi(k-1)),\ D(t, \xi(k-1))\}$

Find sampling time $\xi(k)$ with minimum $D_{t,j}$, i.e., e_t^*

$e_t^* = \min\left\{D_{t,\xi(k)}\right\}$

Find the optimal path $\widehat{\mathbf{f}}_t^{*T}$

If $t < \gamma$ **then**

Compute optimal path $\mathbf{f}_{1,t}^{*T}$ for $\zeta_{1,t}$ at the internal $[1, t]$

else

Compute optimal path $\mathbf{f}_{t-\gamma+1,t}^{*T}$ for $\zeta_{t-\gamma+1,t}$ at the interval $[t - \gamma + 1, t]$

end

Step 3: Integrate optimal path into final solution $\widehat{\mathbf{f}}_t^T$

If $t < \gamma$ **then**

Integrate $w(t, \xi(k))$ of $\mathbf{f}_{1,t}^{*T}$ from $\zeta_{1,t}$

$\widehat{\mathbf{f}}_t^T = \widehat{\mathbf{f}}_t^T \cup \mathbf{f}_{1,t}^{*T}$

else

Integrate $w(t, \xi(k))$ of $\mathbf{f}_{t-\gamma+1,t}^{*T}$ from $\zeta_{t-\gamma+1,t}$ for the $(t - \gamma + 1)$th time

$\widehat{\mathbf{f}}_t^T = \widehat{\mathbf{f}}_{t-1}^T \cup \mathbf{f}_{t-\gamma+1,t}^{*T}$

end

Step 4: Slide forward current window ζ

If $t < \gamma$ **then**

No change in window, i.e., $\zeta_{1,\gamma}$

else

Move current window $\zeta_{t-\gamma+1,t}$ one position forward to $\zeta_{t-\gamma+2,t+1}$

Set the start point of $\zeta_{t-\gamma+2,t+1}$ to the last point $w(t, \xi(k))$ of $\zeta_{t-\gamma+1,t}$

end

Step 5: Updating boundaries

If $t > \max(K_i), i = 1, \ldots I$ **then**

$u_t = u_{t-1}$

$l_t = l_{t-1}$

else if $e_t^* = u_t$ **then**

$u_i = \min(u_i + 2, K_{ref}), i = t + 1, \ldots, \max(K_i)$

else if $e_t^* = l_t$ **then**

$$l_z = \begin{cases} l_z - 1 & z = \{t+1, t+2\} \\ l_z - (l_{t+2} - l_t) & z = t + 3, \ldots, \max(K_i) \end{cases}$$

end

distance occurs, is assessed, and the optimal path is subsequently calculated, which is equivalent to Kassidas et al.'s online implementation. However, none of the pairs $[\iota(k)\xi(k)]$ from the optimal path $\mathbf{f}^{*T}$ within the first window $\zeta_{1,3}$ is saved to the final solution $\widehat{\mathbf{f}}^T$, and consequently, no synchronization is carried out (see blue warping profile in Figure 4.14a). When the measurements belonging to the 4th sampling time are collected, the synchronization window must be moved one position forward, leading to the window $\zeta_{2,4}$ (see Figure 4.14b). At this point, the pairs $[\iota(k)\xi(k)]$ (green line in Figure 4.14b) corresponding to the first ongoing sampling time are saved to the final solution $\widehat{\mathbf{f}}^T$, and the first synchronized batch samples are then available. Note that, following this approach, the synchronization is lagged $\gamma = 3$ sampling units. Again, the end point e^* is assessed and the optimal path $\mathbf{f}_{2,4}^T$ within the window $\zeta_{2,4}$ is calculated (blue line in Figure 4.14b). Afterward, the lower boundary is updated since the end point e^* is just located in one of the points belonging to the mentioned boundary (see dotted lower limit in Figure 4.14b). The procedure to perform the batch synchronization at the 4th sampling time is repeated in the consecutive sampling times (Figure 4.14c–f). Recall that the estimation of the optimal path $\widehat{\mathbf{f}}^T$ from the RGTW algorithm is optimized window-to-window, discarding the wrong synchronization steps (magenta lines shown in Figure 4.14c–f).

Upon finalization of the synchronization of the ongoing batch, there will be synchronized samples at the $(t - \gamma + 1)$th batch time. Depending on whether the algorithm compresses or expands the samples, an update of the last monitored sample – an average of previous samples that matches with the same reference point is calculated as a new measurement – or new values – replicates of the previous synchronized sample – will be computed. As a result, an incomplete batch in the reference timescale is available for monitoring purposes. Throughout this book, we chose the TSR[4, 5] method to impute unknown samples, which is a simple and efficient method to estimate missing data.

4.2.4.1 Enhanced Global Constraints

The use of global constraints avoids large deviations from the diagonal path, preventing similar features in different process phases from being matched. For a feasible search, the band size M has to be equal or greater than the absolute value between the reference and sample batch duration. Typically, the band size M is set to the maximum, median, or average batch duration from NOC batches. The advantage of this strategy is that a large search space is available to obtain the optimal synchronization solution. However, the actual variability of the process in terms of duration is not considered, which may cause similar features to be incorrectly synchronized over time. This drawback may affect the synchronization to a larger extent if notable differences in batch length are found stage-to-stage or phase-to-phase across batches. To overcome these setbacks, the search space should be defined based on the consistency from batch to batch, i.e., boundaries ought to be narrowed at time periods with less deviation from the diagonal path.

In light of the inefficiencies observed with traditional global constraints, the authors propose a band boundary that is more suitable for batch process monitoring. The first step is to synchronize these trajectories with the DTW algorithm without global constraints. The main objective is to find the optimal path across all batches

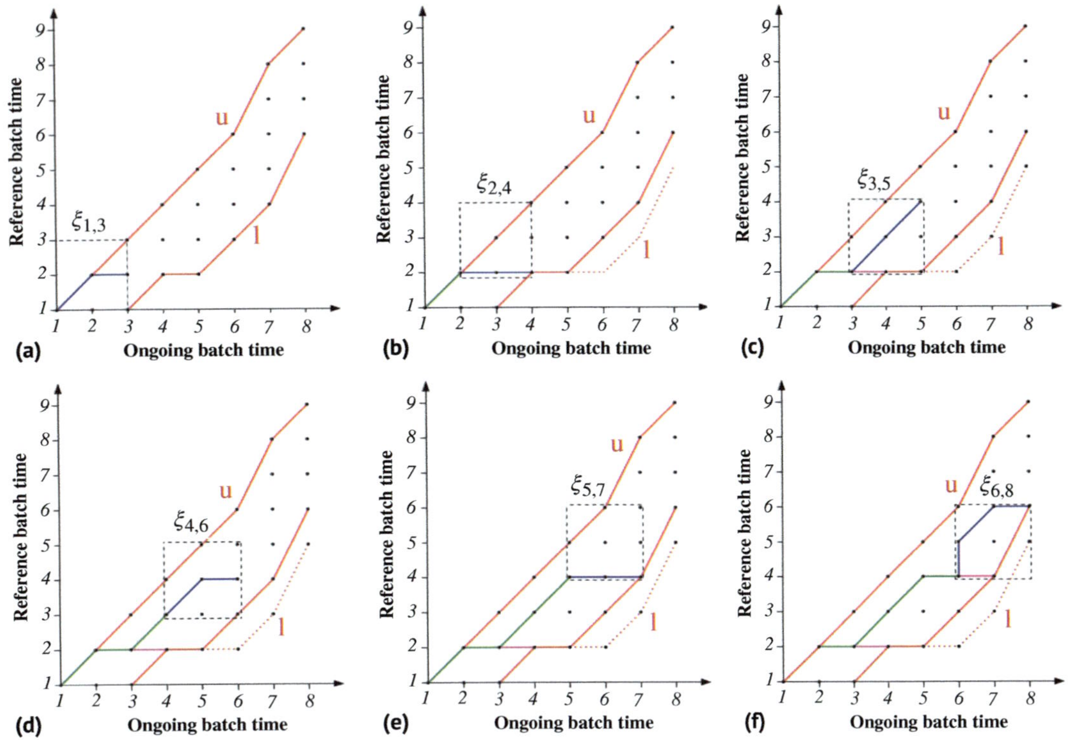

Figure 4.14 Performance of the online synchronization of an ongoing batch from the 1st up to the 8th sampling times in chronological order (from a to f), using the RGTW algorithm with a window width γ equal to 3 units. Red solid lines represent the boundaries that restrict the optimal path search and the red dashed line represents the lower boundary update. The final optimal path $\widehat{\mathbf{f}}_t^T$ and the optimal path $\widehat{\mathbf{f}}_{t-\gamma+1}$ found in the window $\zeta_{t-\gamma+1}$ are plotted by green and blue solid lines, respectively. Magenta solid lines are path segments that are not considered in the final solution, and they are deemed as incorrect synchronization.

at the risk of matching similar features belonging to different phases. In case nonoptimal global constraints are used, there would be no guarantee that the optimal path could be derived. As a result, a matrix $\mathbf{F}\,(I \times K_{w_i})$ containing all optimal paths is generated, where I is the number of NOC batches and K_{w_I} is the number of warping indexes for the ith batch. From all optimal paths, new boundaries are defined as:

$$\begin{aligned} \hat{u}_t &= \max(\mathbf{F}_{1,w(t,\xi(k_1))}, \ldots, \mathbf{F}_{I,w(t,\xi(k_I))}) \\ \hat{l}_t &= \min(\mathbf{F}_{1,w(t,\xi(k_1))}, \ldots, \mathbf{F}_{I,w(t,\xi(k_I))}) \\ t &= 1, \ldots, \max(K_i) \text{ and } i = 1, \ldots, I \end{aligned} \tag{4.17}$$

where $\hat{u}_t$ and $\hat{l}_t$ are the upper and lower boundaries at the tth sampling time, respectively. These boundaries generate an envelope that includes all possible ways in which the optimal path may go along to synchronize multivariate batch trajectories based on the synchronization of NOC batches (see Figure 4.15).

Note that the synchronization should be reevaluated by using a well-defined band global constraint, provided some process events temporally spaced are incorrectly aligned. The warping information from the first synchronization should be used to define a band that enables a more restricted search of the optimal solution in periods with less variability and a less restricted search in periods with more variability in the process pace. This constraint would allow the synchronization algorithm to find the optimal solution and avoid that similar features belonging to different process phases are aligned. For instance, assume that for the simulated example shown in Figure 4.15, wrong alignments were obtained in the first half of the batch runs, where the differences in pace and length are smaller than in the second half of the batch

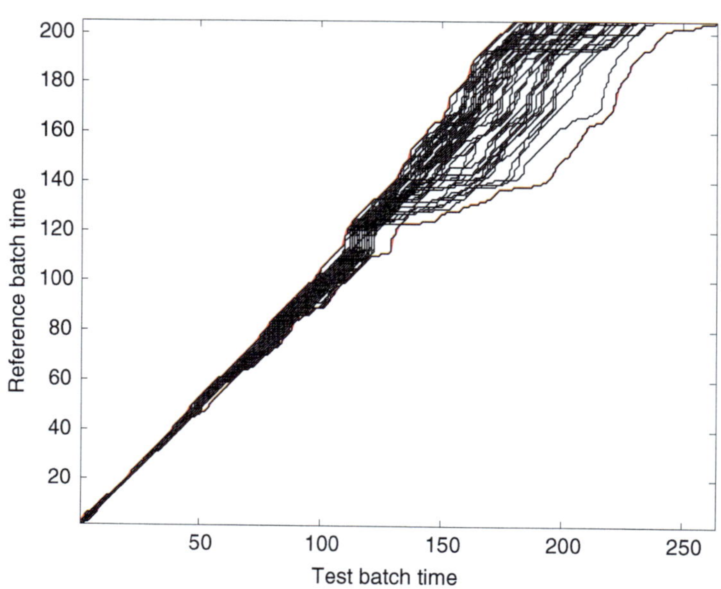

Figure 4.15 Set of optimal paths found by applying the offline DTW algorithm without global constraints. The red lines are the minimum ($\hat{l}_t$) and maximum ($\hat{u}_t$) values at each sampling time, which define an envelope.

runs. These misalignments would be reflected in the warping profiles in the form of large vertical transitions. To avoid the occurrence of wrong matches, two band global constraints could be defined: a band for the first 100 sampling times whose band size is $M_1 = 15$, and a second band for the remaining part of the run whose band size is $M_2 = 60$. Note that these bands are not so restrictive as the ones imposed from the warping information shown in Figure 4.15, but are restrictive enough to limit the matching of common features from different periods between the reference and the sample batch. Once an optimal synchronization is derived, the upper and lower boundaries can be estimated from the warping profiles as explained above.

4.2.4.2 Cross-validation for the Estimation of the RGTW Parameters

In order to apply the RGTW algorithm, the width γ of the synchronization window ζ and the boundaries, must be set up by following a cross-validation procedure.[67] This procedure outlined in Figure 4.16 is further described below.

First, the matrix with I optimal paths $\mathbf{F}_{DTW}$ $(I \times K_{ref})$ is achieved after applying the offline DTW algorithm without global constraints to the raw batch data. Later on, the raw batch trajectories are split into a training set (train) and test set (test), I times in the loop. This step is carried out following the leave-one-out procedure. Every number of window widths γ is used to synchronize the train set in each iteration. As a result, a matrix with $I - 1$ suboptimal paths and window width γ $\mathbf{F}_{RGTW,\gamma}$ is obtained. Once the upper and lower boundaries are defined from those $I - 1$ paths, the test batch is synchronized using the online approach, leading to a suboptimal path $\mathbf{f}^T_{i,RGTW}$, where $i = 1, \ldots, I$. This procedure is repeated until all I batches are synchronized over all window widths. The Pearson's correlation coefficient $(r(\mathbf{f}_{i,DTW}, \mathbf{f}_{i,RGTW}))$ between the suboptimal paths found in the loop for each window width γ and the optimal path from the offline synchronization for all NOC batches is assessed.

Once the above procedure has been executed, a study of the performance of the RGTW-based real-time synchronization as a function of the different window widths

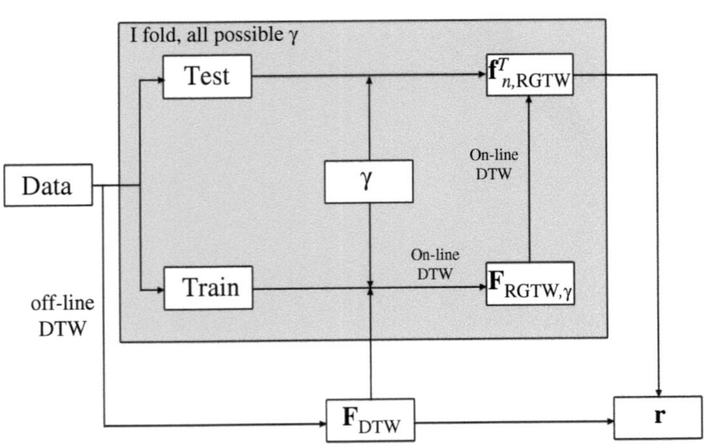

Figure 4.16 Iterative procedure to assess the performance of the RGTW algorithm using different window width γ.

γ studied is necessary. The most adequate procedure is to perform an aalysis of variance (ANOVA) on the Fisher Z-transformed correlation coefficients[2] to determine whether there are statistically significant differences among the window widths under study or not, following the idea published by González et al.[67] If no statistically significant differences are found among the window widths, the smallest width will be selected to improve computational efficiency and enable non-delayed synchronization, which might otherwise be hindered by larger window widths at the beginning of the batch. Conversely, if significant differences are observed, the selection will represent a trade-off between synchronization accuracy – high Fisher Z-transformed correlation – and short synchronization delay – small window widths. In principle, the optimal choice will be the shortest window width that yields a statistically significant improvement in correlation compared to the initial window widths under consideration. When the window width to be used for the RGTW has been selected, the upper and lower boundaries are estimated as the envelope of the warping paths obtained in the cross-validation procedure for such γ.

An assumption in the estimation and optimization of the synchronization parameters for online applications is that the data used is representative of the normal operating conditions of the process under study. However, there could be significant disturbances, such as composition variations, manufacturing campaigns, and seasonal changes that were not reflected in the calibration dataset, but which might appear while monitoring the process. In these cases, the meta-parameters of the subsequent multivariate model built for real-time monitoring should be adapted or even subjected to a new modeling, depending on the nature of the disturbances or conditions, and on the changes of the correlation structure. Likewise, an update of the synchronization parameters (weights of the process variables, sliding window, and boundaries) should be performed to accommodate new patterns of asynchronisms and process paces. If the adaptation of the synchronization parameters is neglected, it may add false features in the trajectories or even misalign key process events. The inaccurate synchronization of the batch trajectories would likely lead to an increase of the false alarms in the control charts. To carry out this adaptation, apart from recalculating the weights given to the process variables to perform the synchronization with new historical batches, the RGTW parameters should be tuned again by reevaluating the cross-validation procedure. Nonetheless, in cases where incoming batches are considered as NOC by the statistical model, assuming that the warping information derived from synchronization is considered as another process variable, the reestimation of the synchronization parameters would not be required.

Example 4.2: Batch synchronization with the RGTW algorithm

The MATLAB live script "ex4_2.mlx" provides the set of MATLAB commands to illustrate how end-of-batch or post-mortem online synchronization can be performed. This procedure is necessary in order to ensure that models are applicable

---

2 Fisher transformation is an elementary transcendental function called the inverse hyperbolic tangent function. This transformation on the Pearson's correlation coefficient is required to approximate it to a normal distribution.

in real time, for which the synchronization must be conducted under the assumption that new data is continuously arriving, without prior knowledge of the remaining batch data. This will ensure alignment with the modeling strategy coded in the MVBatch toolbox, which calibrates the model and resulting monitoring system in such a way that the multivariate statistics are generated in an online manner, assuming that the rest of the batch at a certain sampling time is unknown.

In this example, we load a different dataset of the fermentation of the *Saccharomyces cerevisiae*, and proceed with the DTW synchronization, selecting the batch whose length is the closest to the median as a reference, similarly to the previous example. In addition, none of the process variables will be constrained, and all of them will be used in the computation of the weights. For the computation of the weights, we will execute the same synchronizations as in the previous examples and, finally, calculate the geometric mean of the weights derived from completing the batch synchronization following the Kassidas et al.'s and Ramaker et al.'s approaches.

Prior to the synchronization of the batch trajectories in an online manner using the RGTW algorithm, the width of the window to be used by the algorithm must be set – in the example we use a value of zetacv = 4 – and validated by the cross validation procedure explained in Section 4.2.4.2. For this purpose, the Pearson's correlation coefficients between the suboptimal paths from the offline DTW-based synchronization are calculated.

```
[rgtwWARPS,rgtw_r] = cvBands(raw_data_scaled,diag(
   weights_geom),band,...
    reference_batch,warping_profiles_raw_geom',zetacv);
```

This allows us to determine the optimal window width that minimizes the lack of fit caused by the greedy strategy. The next step is to perform an ANOVA on the Fisher Z-transformed correlation coefficients for the different window widths.

```
[~,~,st] = anova1(atan(rgtw_r'));
```

An ANOVA was performed on the Fisher Z-transformed correlation coefficients for the different window widths, showing statistically significant differences (p-value < 0.05). The least significant difference (LSD) intervals displayed in Figure 4.17 show that correlation is statistically higher for window width $\gamma = 4$, but no statistically significant differences were found for windows with widths between $\gamma = 1$ and $\gamma = 2$. Note, however, that in this example, $\gamma = 3$ is the shortest window width that yields a statistically significant improvement in correlation compared to $\gamma = 1$ and $\gamma = 2$. Therefore, we advocate for using a window width $\gamma = 3$ as a compromise between computational complexity and monitoring performance for subsequent synchronizations.

Once the width of the synchronization window jointly with the band (global constraints) has been optimized, the last step is to proceed with the end-of-batch online synchronization using the RGTW algorithm. An example of the execution of this synchronization with a single batch is illustrated below.

```
[synchronized_batch_rgtw,warp] = onSyn(raw_data,Bref,band,
   diag(weights_geom),zeta,rng);
```

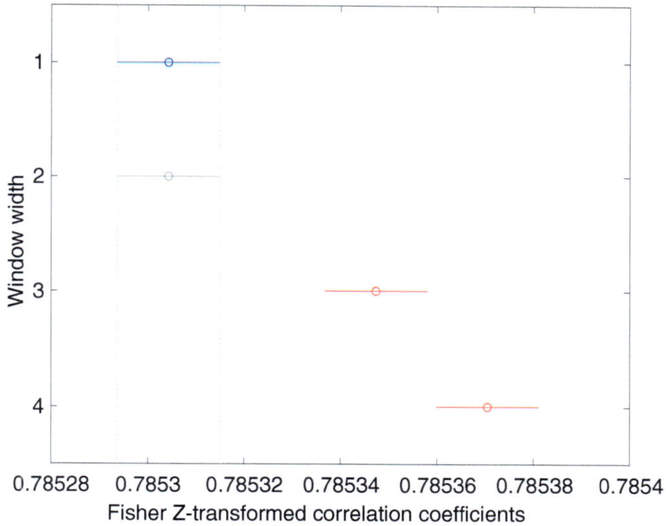

Figure 4.17 LSD intervals for the Fisher Z-transformed correlation coefficients for the different window widths γ.

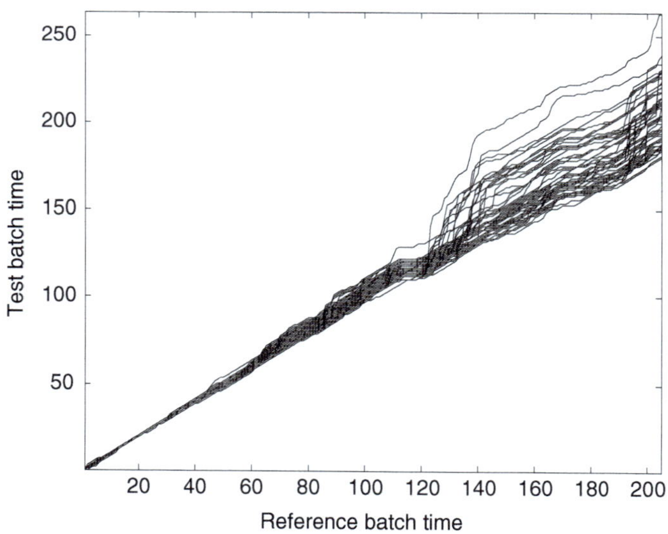

Figure 4.18 Distortion of the batch time over all batches (warping information) after applying the RGTW algorithm.

where the inputs are the raw batch data corresponding to the batch to synchronize, the autoscaled reference batch data to use as a reference for synchronization, the matrix of weights, the width of the synchronization window, and the array of ranges associated with all the process variables.

Figure 4.18 shows the warping profiles obtained from the validation procedure using $\gamma = 3$. As graphically illustrated, the warping information is smooth for all batches from the start of the batch to the 50th sampling time. This means that almost no compressions or expansions were necessary to synchronize all batches. In contrast, a greater number of vertical and horizontal transitions were required to align the batches from the 50th to the 120th sampling time point. Specifically, a series of expansions (horizontal paths) can be observed from the 105th to the 120th sampling time. One reason for this fact may be the phase transition in the process. From this point, an important number of compressions (vertical paths) were needed because many batches have longer duration than the reference batch.

Warping information can be used to improve troubleshooting of problems that may be masked by synchronization. For that purpose, we can add the warping information expressed as a function of the reference batch directly into the synchronized data two-way array, along with the other variables. However, caution needs to be taken when the warping information is taken into account in LVMs. The inclusion of warping information as a new variable may not be completely correlated with any of the process variables, and consequently, it may have a negligible effect on the model. In such cases, this variable should be considerably weighted in order to avoid omitting the process variability associated with the corresponding process variable in batch data modeling.

4.2.5 Multisynchro

The presence of multiple asynchronisms in the batch trajectories also poses a threat to bilinear modeling. We can think of at least four different types of asynchronism: (i) batches with equal duration but key process events not overlapping at the same sampling time in all batches (class I asynchronism), (ii) batches with different duration and process pace (class II asynchronism), (iii) batches with different duration due to incompletion of the last process stage, irrespective of whether the process pace is the same or not across batches (class III asynchronism), and (iv) batches with different duration due to delay in the start but batch trajectories showing the same evolution pace after (class IV asynchronism). Traditional synchronization approaches cope with asynchronism issues by applying a single synchronization procedure to all batches without considering the nature of asynchronism of each batch. In a context of multiple asynchronisms, applying the same synchronization procedure may distort the original trajectories and decrease the signal-to-noise ratio, adversely impacting the original correlations of the process variables over time.

González et al.[66] proposed a novel synchronization approach named Multisynchro to deal with batches affected by inadvertent anomalies and multiple asynchronisms. The new approach uses the valuable information on the process pace of each batch derived from DTW/RGTW-based synchronization (the so-called warping information) for two purposes: (i) detecting the type of asynchronism of each particular batch and (ii) deploying the appropriate synchronization procedure based on the nature of asynchronisms. The new approach also includes a procedure that performs anomaly detection and batch synchronization on non-shifted and complete batches in an iterative way. In addition, a discussion on the consequences of common practices in synchronization of complex scenarios of asynchronisms is provided. More precisely, the Multisynchro algorithm consists of a high- and low-level routine (see Figure 4.19). The high-level routine is aimed at recognizing the different types of asynchronous trajectories for the subsequent batch classification as function of the nature of asynchronism (see Figure 4.19a). The low-level routine is aimed to synchronize the variable trajectories of each one of the batches with a specific procedure based on the type of asynchronism (see Figure 4.19b). The Multisynchro algorithm is further described below.

4.2.5.1 Asynchronism Detection

The high-level routine is divided into two steps (see Figure 4.19a). The first step is devoted to recognizing the different types of asynchronous trajectories, which is carried out by using the warping time profiles derived from a preliminary synchronization as follows:

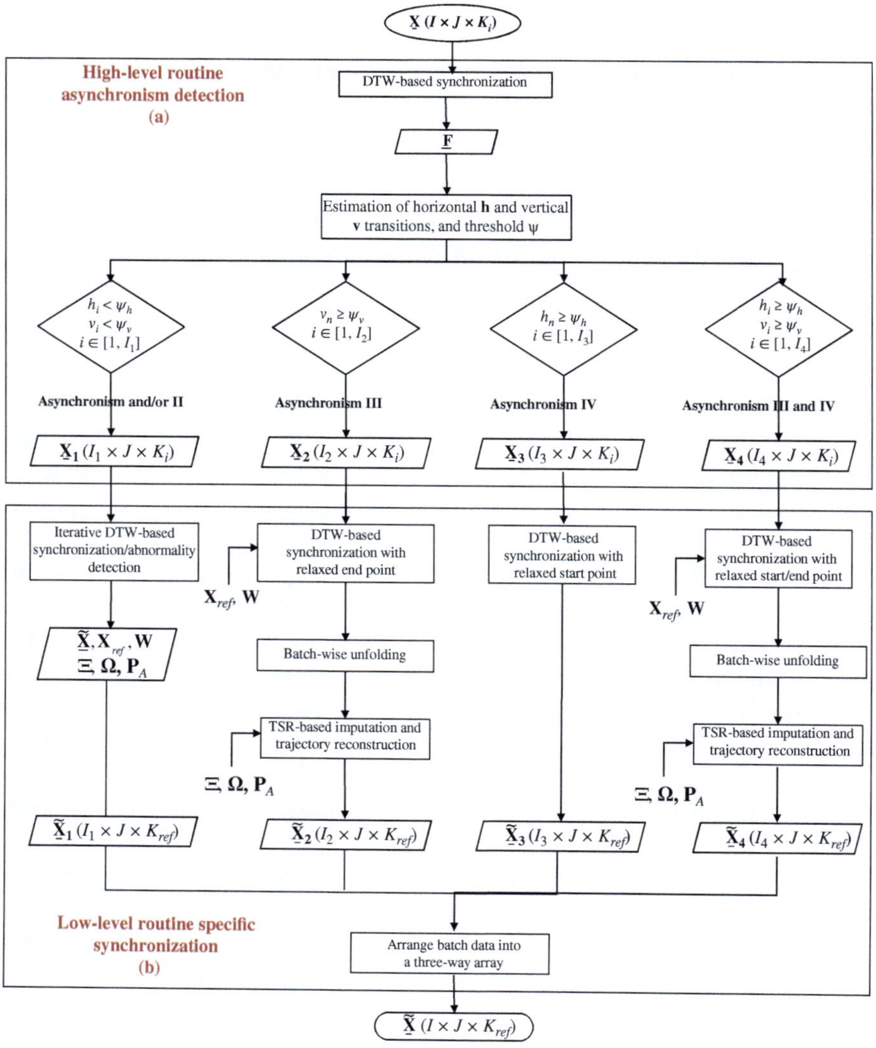

Figure 4.19 Flow diagram of the multisynchro approach composed of the (a) high-level and (b) low-level routines for batch synchronization in scenarios of multiple asynchronisms.

i. Select a complete NOC reference batch **X**_ref from the three-way batch data array $\underline{\mathbf{X}}$.

ii. Synchronize all batches using the DTW algorithm by giving the same weight to the process variables that contain valuable information for synchronization. Those variables that are either showing constant values in most of the production time or discarded beforehand by prior knowledge are constrained in the synchronization with a null weight. The reason certain process variables are given the same importance is to mitigate the distortion of the warping profiles in the presence of different types of asynchronisms. The algorithm returns a three-way synchronized batch data array $\underline{\tilde{\mathbf{X}}}$ and a three-way array $\underline{\mathbf{F}}$ $(I \times 2 \times K_{w_i})$ containing the warping paths for the I batches.

iii. For each warping time profile $\mathbf{f}_i$ from the three-way array $\underline{\mathbf{F}}$:

iii.1 Count the number of consecutive horizontal transitions at the first time period of the ith batch, which denotes the number of compressions carried out by the synchronization algorithm at the start of the ith batch:

$$h_i = \sum_{k=1}^{Kw_i} (\xi(k) = 1)$$

iii.2 Count the number of consecutive vertical transitions at the last time period of the ith batch, which indicates the number of expansions carried out by the synchronization algorithm at the end of the ith batch:

$$v_i = \sum_{k=1}^{Kw_i} (\iota(k) = K_i)$$

The parameterization of patterns on the warping time profiles is used to detect the different types of asynchronisms that might be present in data. If there are batches affected by class I and/or II asynchronism, the resulting warping profiles are expected to show a reasonable combination of horizontal and vertical transitions throughout the batch run. Batches that are influenced by class III asynchronism are associated with warping profiles containing an excessive number of vertical transitions at the last time period of the runs. Batches shifted at the start of the run will show a high number of horizontal transitions at the start of the batch on their corresponding warping profiles, assuming that the reference batch is selected from the set of non-shifted batches (asynchronism IV).

The second step of the high-level routine (see Figure 4.19a) is aimed at classifying each batch by the type of asynchronism and arranging them into different datasets. For this purpose, the features of the warping time profiles obtained from the first step of the high-level routine are used to distinguish among asynchronisms. To determine whether the number of transitions exceeds the normal limit from which a batch can be qualified as affected by class III and/or IV asynchronism, rather than by class I and/or II asynchronism, thresholds ψ_v and ψ_h are derived. These thresholds are calculated as a fraction $\varkappa$ of the interquartile range of both the vertical transitions at the last time period $\mathbf{v}$ and the horizontal transitions at the first time period $\mathbf{h}$ estimated for all the synchronized batches, respectively. Even though $\varkappa$ is a heuristic value dependent on the distribution of the transitions, it is recommended that $\varkappa$ does not exceed 0.5. Also, note that the interquartile range is used as a dispersion statistic due to its robustness to outliers and extreme values, which breaking point is 50% of extreme values in the dataset.

i. Repeat for all batches:

i.1 If the number of compressions h_i and expansions v_i are less than their respective thresholds, arrange the ith batch into the three-way array $\underline{\mathbf{X}}_1$, which contains batches affected by class I and II asynchronisms(see classes I and II in Table 4.1).

i.2 Only if the number of expansions at the end of the batch v_i is greater than or equal to the threshold ψ_v, arrange the ith raw batch into the three-way array $\underline{\mathbf{X}}_2$, which contains batches affected by class III asynchronism (see class III in Table 4.1).

i.3 Only if the number of compressions at the start of the batch h_i is greater than or equal to threshold ψ_h arrange the ith raw batch into the three-way array $\underline{\mathbf{X}}_3$ by class IV asynchronism (see class IV in Table 4.1).

i.4 If the number of compressions h_i and expansions v_i are greater than or equal to their respective thresholds, arrange the ith raw batch into the three-way array $\underline{\mathbf{X}}_4$ by class III and IV asynchronisms (see classes III and IV in Table 4.1).

4.2.5.2 Specific Batch Synchronization

The Multisynchro approach continues the execution synchronizing the different datasets with different types of asynchronism (low-level routine depicted in Figure 4.19b):

i. Synchronize the three-way batch data array $\underline{\mathbf{X}}_1$ containing the batches whose trajectories are neither shifted nor incomplete in such a way that possible anomalies present in batch data do not affect the synchronization quality. First, the iterative synchronization based on the DTW algorithm is applied to the whole dataset. The isolation of the faulty synchronized batches is then conducted with a PCA model built on the assumed NOC synchronized batches. Once done, the NOC batches identified by the model with their raw variables trajectories are again synchronized with the iterative DTW algorithm. These steps are iteratively performed until no more outliers are detected. For a rigorous description of this iterative batch synchronization/anomalies detection procedure, readers are referred to Section 4.2.5.3.

The detection of suspicious batches in $\underline{\mathbf{X}}_1$ is crucial to obtain accurate parameters that are not affected by anomalies for the synchronization of the rest of datasets and for the imputation of missing trajectories. Note that, if the batches of other datasets were synchronized with this procedure, the asynchronisms affecting these batches would negatively impact the synchronization results. The DTW algorithm, without relaxed end point constraints, would create flat profiles when batches are incomplete at their last stage, or might add incorrect features at the start of the batch when batches are shifted. As a consequence,

Table 4.1 Case-based reasoning of the multisynchro algorithm.

Class	Asynchronism illustration	Description
I	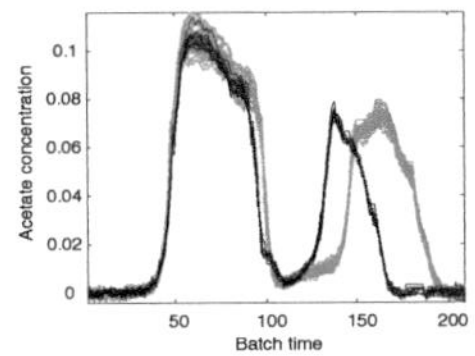	Batches with equal duration but key process events not overlapping at the same sampling time point in all batches are classified as batches affected by class I asynchronism. Batches with data manipulated depicted in black lines will be classified as batches affected by class I asynchronism because no shifts are appreciated at the start of the batches ($h_i < \psi_h$) and completion of the last stage of batches is observed ($\upsilon_i < \psi_\upsilon$), assuming that the reference batch is selected from those batches that are not manipulated (gray batches in the left-hand figure)
II	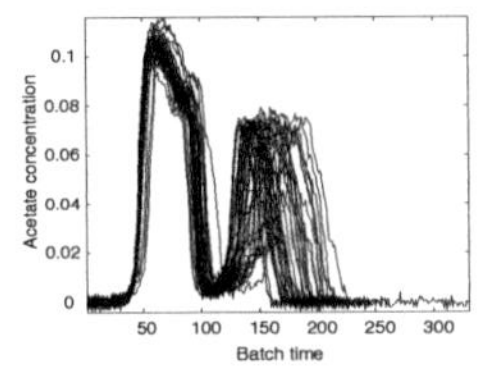	Batches with different duration and process pace are classified as batches affected by class II asynchronism. The left-hand side figure shows different process pace across batches, and as in the previous case, neither shifts ($h_i < \psi_h$) nor incompletion ($\upsilon_i < \psi_\upsilon$) are observed. The difference between class I and II lies in the warping profiles: (i) higher variation in the number of compressions and expansions over the batch run in class II than in class I and (ii) existence of different matching end points in class II due to different lengths in all batches, which is a feature not observed in class I
III	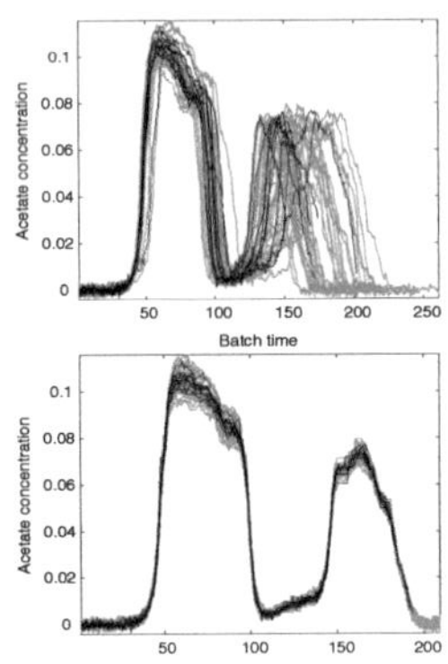	Batches with different durations due to partial completion of the last process stage, irrespective of pace differences across batches, are classified as batches affected by class III asynchronism. Manipulated batches shown in black lines in the left- and right-hand figures will be classified as class III asynchronism batches because $\upsilon_i \geq \psi_\upsilon$, as long as the reference batch is selected from the set of non-manipulated batches (gray batches in the figures). Note, however, that their warping profiles will show differences since the process pace is different over all batches in the left-hand figure, in contrast to those depicted in the right-hand figure. Regarding the non-manipulated batches depicted by gray lines, those shown in the left-hand figure will be treated as class II due to the difference in pace. Those depicted in the right-hand figure may be treated as either class I or class II asynchronous batches, depending on whether these batches have different durations in comparison to the reference batch
IV	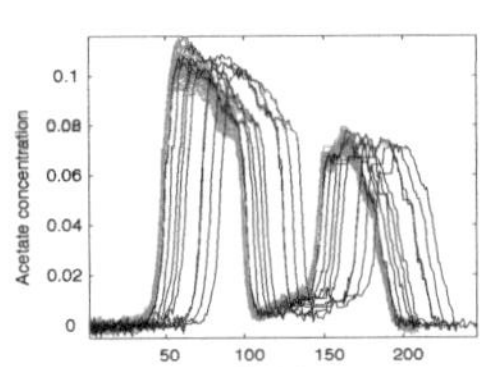	Batches with different durations due to delay in the start but trajectories showing the same evolution pace thereafter are classified as batches affected by class IV asynchronism. Manipulated batches depicted in black lines will be classified as batches affected by asynchronism IV because $h_i \geq \psi_h$, assuming the reference batch is one of the non-manipulated batches (gray batches). Batches that are not manipulated will be classified as batches affected by class I and/or II asynchronisms because $h_i < \psi_h$ and $\upsilon_i < \psi_\upsilon$, and the assignment will depend on whether the duration is the same or not in relation to the reference batch

the weights of the process variables, the reference batch estimated as the average trajectory, and ultimately, the alignment of the process events would be severely perturbed. The poor synchronization quality would compromise the correct segregation of the batches in NOC and faulty, and eventually, the overall synchronization. This knock-on effect would inevitably lead to the estimation of inaccurate preprocessing parameters and, most importantly, distort the actual correlation structure.

At the end of the synchronization of $\underline{\mathbf{X}}_1$, the procedure returns the matrices of average trajectories Ξ and standard deviation trajectories $\boldsymbol{\Omega}$, the weighting matrix $\mathbf{W}$, the loading vector $\mathbf{P}_A$ obtained from the PCA-based modeling, and the three-way array $\underline{\widetilde{\mathbf{X}}}$ that arranges the synchronized NOC and faulty batches.

ii. Synchronize the three-way batch data array $\underline{\mathbf{X}}_2$ containing only incomplete batches in such a way that their last point is well synchronized with the best matching sampling time of the reference. This type of synchronization is needed to prevent the incomplete batch trajectories from being misaligned and to avoid the addition of flat profiles that break the correlation structure. For this purpose, the DTW algorithm with the relaxed end point constraint using those parameters estimated in the iterative synchronization can be applied. This version of the DTW algorithm synchronizes batches against a segment of the reference batch limited by the first point and the best matching end point e^* instead of the reference as a whole. The algorithm returns the batch trajectories synchronized till the best end point of each batch $\underline{\widetilde{\mathbf{X}}}_2$. The missing part of each batch is then imputed using the correlation structure captured by the multivariate model obtained from the iterative DTW-based synchronization/anomaly detection procedure through the application of the TSR method.[4] The procedure returns the three-way array $\underline{\widetilde{\mathbf{X}}}_2$ containing the synchronized batch trajectories.

iii. Synchronize the three-way batch data array $\underline{\mathbf{X}}_3$ using the DTW algorithm with the relaxed starting point constraint and the parameters calculated in the iterative synchronization. This version of the DTW algorithm synchronizes segments of batches against a reference batch. The segments are limited by the best matching starting point s^* of each batch with the first point of the reference, and their last point. In other words, this procedure truncates the shifted batches to the best matching point with the reference batch to mitigate the risk of the iterative synchronization introducing artifacts when the first samples are shrunk or expanded. The presence of noise or even unexpected landmarks until the first best matching point with the reference batch might add undesired features in the synchronized trajectories. The procedure returns the three-way array $\underline{\widetilde{\mathbf{X}}}_3$ containing the synchronized batch trajectories.

iv. Synchronize the three-way batch data array $\underline{\mathbf{X}}_4$ using the DTW algorithm with the relaxed starting and end point constraint using those parameters estimated in the iterative synchronization. The procedure returns a three-way array $\underline{\widetilde{\mathbf{X}}}_4$ containing the synchronized batch trajectories.

The application of different nonlinear warping functions with different end point constraints might alter the covariance matrix if the asynchronisms are not treated in accordance with their nature. The synchronization of incomplete batches based on Multisynchro does not necessarily imply a modification of the covariance structure because the synchronization obtained is accurate enough from the start of the batch till the last sampling time available. The remaining part is imputed based on the correlation structure of NOC data. If there is a breakage of the correlation structure, it might be most likely caused by an abnormal situation that occurred over the batch. However, the inclusion of the warping information as a new variable in the dataset might help assess whether the breakage of the correlation is caused by the alignment itself or by an unexpected event in the process.

After synchronizing batch data using Multisynchro, all the resulting submatrices need to be merged into an array $\underline{\tilde{\mathbf{X}}}\,(I \times J \times K_{\mathrm{ref}})$ for subsequent bilinear process modeling. In addition, the warping profiles obtained in each specific synchronization are added as a new variable into the synchronized three-way array $\underline{\tilde{\mathbf{X}}}$ for monitoring purposes. The inclusion of synchronized faulty batches in the calibration dataset does not mean that these batches will be considered in the final model for monitoring. If these batches are eventually abnormal, the model built in the next step of the modeling cycle will spot them as such, and hence, after investigation, the batches will be eventually removed from the dataset.

The real-time application of the Multisynchro approach is straightforwardly done by using the RGTW algorithm instead of the DTW algorithm. For offline applications, DTW is preferred since it provides the optimal global solution. However, if the main goal is to design a monitoring scheme for real-time application, the RGTW algorithm is suggested.

4.2.5.3 Iterative Batch Synchronization and Anomaly Detection Procedure

Batch synchronization needs to be implemented, taking into account the possible presence of anomalies in batch data. The existence of faulty batches in the calibration dataset may result in inappropriate synchronizations since possible artifacts may be introduced due to anomalies. For instance, batch trajectories that break the correlation structure usually contain different shapes in comparison to batch trajectories run under NOC. Anomalies may affect the estimation of the weight matrix $\mathbf{W}$ and the synchronization quality, leading to synchronized batch trajectories with artificial shapes at different time periods. To overcome this challenge, an iterative synchronization/anomalies detection procedure is further described.[66] The aim of this procedure is to synchronize each batch with a reference batch and with each other, in such a way that suspicious batches do not affect the synchronization quality. The main steps of the algorithm are the following (see Figure 4.20):

i. Synchronize all the batches contained in the starting three-way array $\underline{\mathbf{X}}$ using the DTW algorithm. For this purpose, select a reference batch $\mathbf{X}_{\mathrm{ref}}$ and a criterion to weight the process variables. The algorithm returns the synchronized three-way batch data array $\underline{\tilde{\mathbf{X}}}\,(I \times J \times K_{\mathrm{ref}})$ and the weight matrix $\mathbf{W}$.

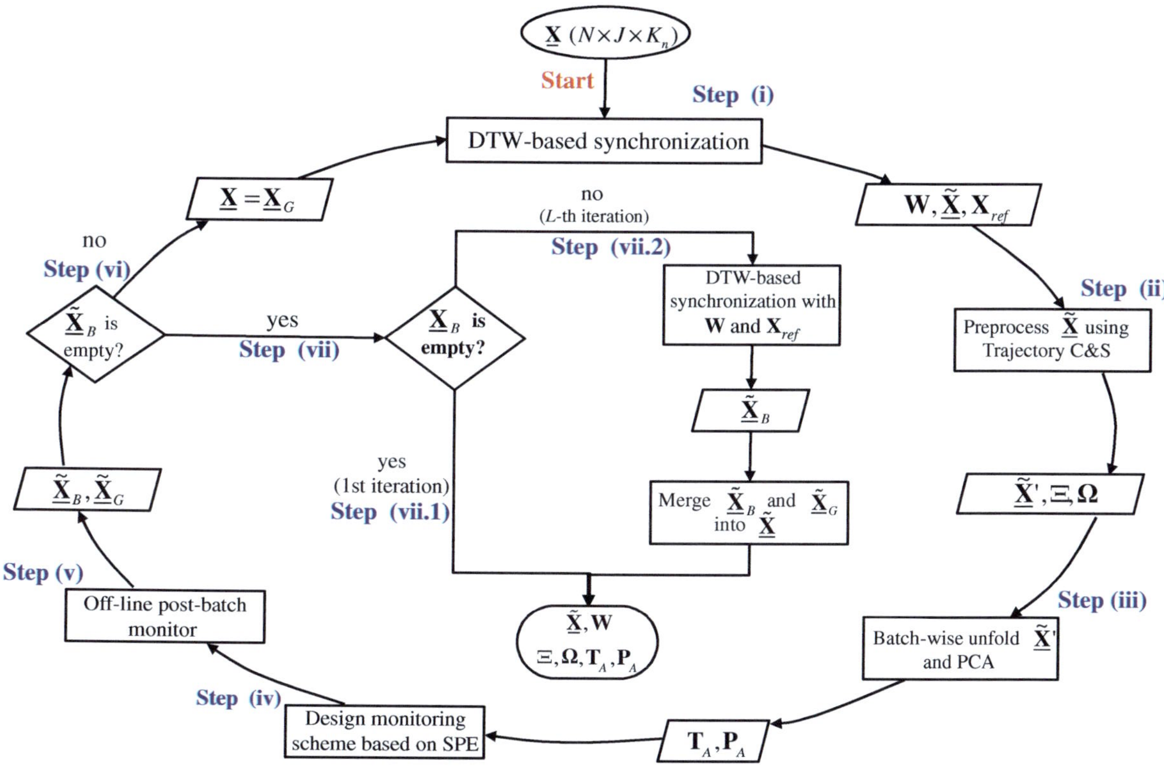

Figure 4.20 Flow diagram of the iterative batch synchronization/anomalies detection procedure. Note that $\tilde{\underline{\mathbf{X}}}_B$ is the three-way array containing the synchronized faulty batches isolated at the *l*th iteration, whereas $\underline{\mathbf{X}}_B$ is the three-way array containing all the raw faulty batches isolated in the *L* iterations of the iterative procedure.

ii. Preprocess batch data by trajectory centering and scaling using the estimated matrices of averages $\Xi\,(K_{\text{ref}} \times J)$ and standard deviations $\boldsymbol{\Omega}\,(K_{\text{ref}} \times J)$[3] to obtain the three-way data array $\underline{\widetilde{\mathbf{X}}}$.

iii. Fit a PCA model on the batch-wise unfolded and preprocessed two-way array $\widetilde{\mathbf{X}}$ satisfying the following equation: $\widetilde{\mathbf{X}} = \mathbf{T}_A \cdot \mathbf{P}_A^T + \mathbf{E}$, where A is the number of PCs extracted.[4]

iv. Design a control chart based on the squared prediction error (SPE) statistic, along with the control limit UCL(SPE)$_\alpha$ estimated from the synchronized calibration batch data at (1-α) confidence limit; see Chapter 7 for details in their computation.

v. Offline post-batch monitor all the synchronized calibration batches for fault detection.

v.1 Compute the SPE statistic for each batch and sort out the corresponding values in ascending order.

v.2 Calculate the acceptable number of batches R that can exceed the control limits at $(1 - \alpha)$ confidence level by chance as α times the number of calibration batches.

v.3 If the number of batches exceeding UCL(SPE)$_\alpha$, I_f, is greater than R, the first $B_l = I_f - R$ synchronized batches with the highest SPE values are treated as faulty batches. In any case, those batches whose SPE values are beyond λ times UCL(SPE)$_\alpha$ are also considered as faulty. To isolate these B_l faulty batches for subsequent synchronization different from that performed on NOC batches, remove them from the dataset and add their respective original batch trajectories into the three-way array $\underline{\mathbf{X}}_B$ $(B_L \times J \times K_b)$, which contains the B_L faulty batches isolated in previous iterations.

v.4 The remaining batches are considered as NOC, and their trajectories are arranged into the three-way array $\underline{\widetilde{\mathbf{X}}}_G$ $(G \times J \times K_{\text{ref}})$.

vi. If one or more batches were detected as abnormal in the offline post-batch monitoring at the lth iteration, compute the repeat loop (i)-(v) with the new calibration batch data array $\underline{\mathbf{X}} = \underline{\mathbf{X}}_G$, where $\underline{\mathbf{X}}_G$ is a $(G \times J \times K_g)$ array containing the raw batch trajectories of G NOC batches.

vii. If no batch was detected as abnormal in the offline post-batch monitoring at the lth iteration, synchronize the faulty batches and merge the datasets.

vii.1 If no batch was detected as abnormal in the first iteration, the iterative procedure terminates.

vii.2 If some batches were detected as abnormal in the offline post-batch monitoring after L iterations, synchronize each faulty batch $\mathbf{X}_b$ from the three-way faulty batch array $\underline{\mathbf{X}}_B$. For this purpose, the DTW algorithm is applied using the reference batch $\mathbf{X}_{\text{ref}}$ and the weighting matrix $\mathbf{W}$

---

3 This preprocessing approach, which is introduced in Chapter 5, is selected due to its suitability for batch process modeling and monitoring.[64, 93]

4 The objective of building a PCA model is to design a monitoring scheme for fault detection. In process monitoring, the focus is on the distributions of latent variables and residuals, which are those used to estimate the control limits for incoming data. This should be taken into consideration to select A.[24]

that were assessed in the NOC batch synchronization in step (i) at the last iteration. Once the synchronized three-way faulty batch array $\underline{\widetilde{\mathbf{X}}}_B$ is available, merge it with the three-way array of NOC batches $\underline{\widetilde{\mathbf{X}}}_G$ into the three-way array $\underline{\widetilde{\mathbf{X}}}$.

As output, the iterative batch synchronization/anomalies detection procedure returns: the three-way synchronized batch data array $\underline{\widetilde{\mathbf{X}}}$, the reference batch $\mathbf{X}_{\text{ref}}$; the matrices of average trajectories Ξ and standard deviation trajectories $\boldsymbol{\Omega}$; the weighting matrix $\mathbf{W}$ assessed in the synchronization procedure; and the score $\mathbf{T}_A$ and loading $\mathbf{P}_A$ matrices obtained from the PCA model on the batch-wise unfolded preprocessed matrix at the last iteration.

As a general rule, the results of the synchronization must be thoroughly examined by a domain expert to ensure that the synchronization algorithm does not contain distorted data. For this purpose, the raw and synchronized NOC batches as well as the raw and synchronized faulty batches should be visually inspected.

Example 4.3: Batch synchronization with the Multisynchro algorithm

The objective of this example is to illustrate (i) the performance of the novel Multisynchro approach for batch synchronization in scenarios of multiple asynchronisms and (ii) the effect of inappropriate synchronization on the batch trajectories by comparing ordinary DTW and Multisynchro. Ultimately, the results of this comparison will enable a discussion on the consequences of common practices in the synchronization of complex scenarios involving asynchronisms. For this purpose, a different dataset generated by the simulator of the *Saccharomyces cerevisiae* fermentation process and manipulated to induce four different classes of asynchronisms is used. The complete sequence of MATLAB instructions to generate the figures of merit can be found in the live script "ex4_3.mlx."

In this example, we load the batch dataset containing four different combinations of asynchronisms: (i) Case #1: 30 batches affected by class I asynchronism that contain trajectories of equal duration but with slightly differing paces across batches, and 10 batches affected by class III asynchronism and consisting of a set of batch trajectories of different lengths caused by a partial completion of the last stage of the batch run, with key process events closely overlapping across batches; (ii) Case #2: 30 batches affected by class I asynchronism, and 10 batches with class IV asynchronism, consisting of batches with different durations due to delay in the measurement collection but whose trajectories show identical evolution pace across all batches; (iii) Case #3: 30 batches with class II asynchronism, consisting of trajectories whose process events do not overlap, and whose lengths differ across batches, and 10 batches with class III asynchronism; and (iv) Case #4: two datasets containing 30 and 10 batches, respectively, with the same length and different evolution pace (see Figure 4.21).

The batches with the four different types of asynchronisms are synchronized by using two approaches for comparison purposes: (i) the ordinary DTW algorithm assigning the same weight to all the process variables and (ii) the Multisynchro approach. For the latter, the high-level routine is executed for asynchronism

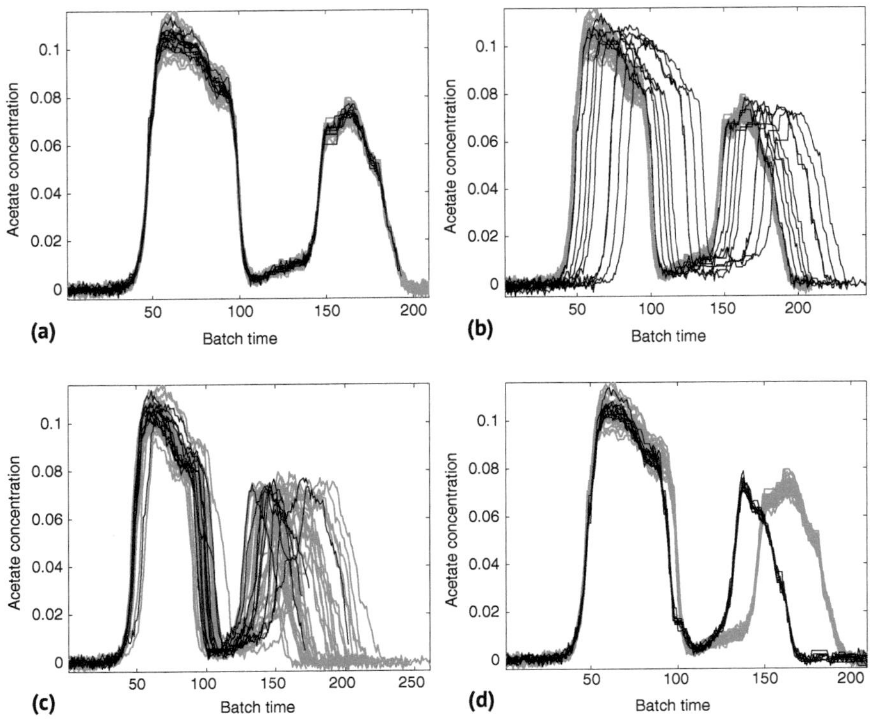

Figure 4.21 Trajectories of the acetate concentration corresponding to 40 NOC batches in four different scenarios of asynchronisms: (a) Case #1: class I (gray lines) and class III (black lines) asynchronisms; (b) Case #2: class I (gray lines) and class IV (black lines) asynchronisms; (c) Case #3: class II (gray lines) and class III (black lines) asynchronisms; and (d) Case #4: class I asynchronism (gray and black lines). The gray lines represent the first dataset of 30 batches and the black lines represent the second dataset of 10 batches. *Source*: Reproduced from Ref.[66], with permission of John Wiley & Sons.

detection, followed by the low-level routine for bespoke synchronization as per the type of asynchronism.

As a result of these synchronizations, the warping profiles are derived, fingerprinting every action taken by the algorithm. This process provides insights into the nature of asynchronism present in batch data.

In Cases #1, #2 and #4, the warping profiles belonging to the 30 out of 40 batches (see gray lines in Figure 4.22a, b, and d, respectively) almost follow the main diagonal. Note that these batches have equal duration and apparently their key process events overlap at the same sampling times across batches (see gray lines in Figure 4.21a, b, and d, respectively). Nonetheless, the slight deviations observed from the diagonal profile denote that, even though most of the batches have equal duration, the main events are not perfectly synchronized. This supports the claim that the batch synchronization is required even when the variable trajectories show the same evolution pace. Regarding Case #3, the warping profiles of the first 30 batches (see gray lines in Figure 4.22c) show a clear variation in the duration of the batches, which is more prominent at the later stages of the process. This type of pattern was expected due to the nature of the asynchronism of these batches. Concerning the

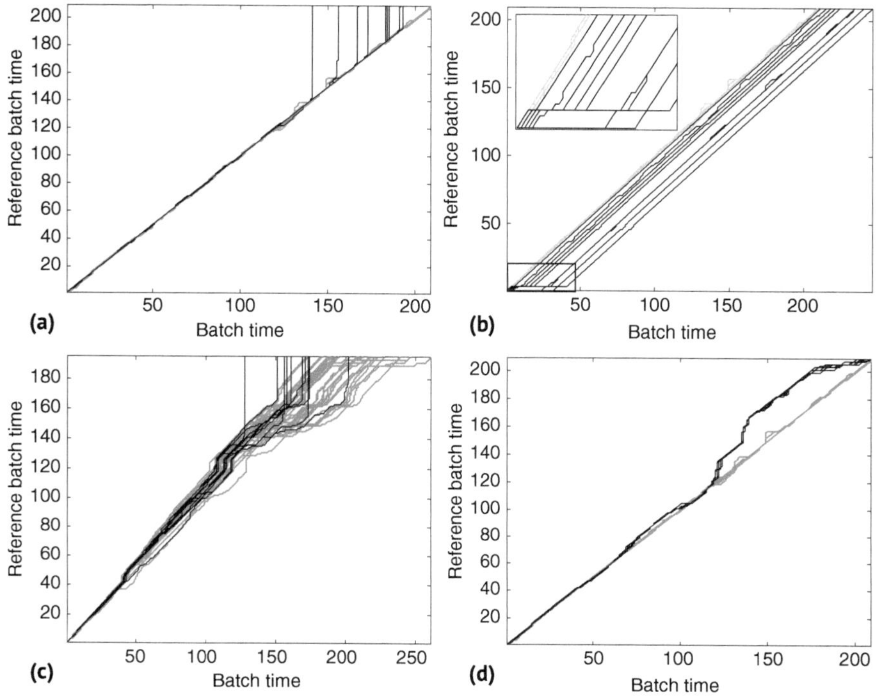

Figure 4.22 Warping information derived from the DTW-based synchronization of the raw batch trajectories for each of the asynchronism scenarios: (a) Case #1: class I (gray lines) and class III (black lines) asynchronisms; (b) Case #2: class I (gray lines) and class IV (black lines) asynchronisms; (c) Case #3: class II (gray lines) and class III (black lines) asynchronisms; and (d) Case #4: class I asynchronism (gray and black lines). The gray lines represent the first dataset of 30 batches, and the black lines represent the second dataset of 10 batches. *Source*: Reproduced from Ref.[66], with permission of John Wiley & Sons.

warping profiles corresponding to the rest of batches (see black lines in Figure 4.22), a different asynchronism pattern is recognized in each case.

In Figure 4.22a, c, 10 out of the 40 warping profiles (black lines) show an excessive number of vertical transitions in comparison to the rest (gray lines). The difference between the two cases is that the raw trajectories in the latter are not synchronized from the start of the batches (see the warping profiles notably deviating from the main diagonal in Figure 4.22c). This pattern is directly related to the presence of batches that were not entirely completed. In practice, such incomplete batches are inadvertently included for batch synchronization, causing severe and undesirable changes in the profiles of the process variables. In Case #2, 10 out of 40 batches (those at which a shift-type asynchronism was introduced) show a diagonal warping profile that is parallel to the main diagonal (see Figure 4.22b). This pattern is typical for cases where similar values of the J process variables are registered at the start of the batch. It leads to an excessive number of horizontal transitions in the synchronization, as observable in the solid black rectangle in Figure 4.22b. The higher the duration of the shift from the start, the higher the number of horizontal transitions in the warping profile. In this case, the DTW algorithm shrinks the corresponding batches at that time interval by averaging the measurements of the J process variables. Finally, the warping profiles related to Case #4 show a deviation from the diagonal, confirming a different process pace within the second half of the run across batches (see Figure 4.22d).

Figure 4.23 shows the resulting warping profiles derived from the low-level multisynchro routine for the four cases of asynchronisms under study. In the cases where the batches are partially incomplete at the end of the batch runs, irrespective of the process pace, the Multisynchro algorithm finds the best matching between their end point and the sampling time of the reference that minimizes the weighted multivariate distance. The relaxation of the end point can be appreciated in the incomplete warping profiles in cases #1 and #3 (see black lines in Figure 4.23a, c). The remaining trajectory of the process variables, including the warping profiles, are imputed by using the variance-covariance matrix of the in-control statistical model obtained from the NOC batches only affected by class I and II asynchronisms. As regard to class IV asynchronism, the resulting warping profiles still show the shift over the main diagonal, although the starting point is the last sampling time of the series of observations that matches the first sampling time of the reference (see black lines in

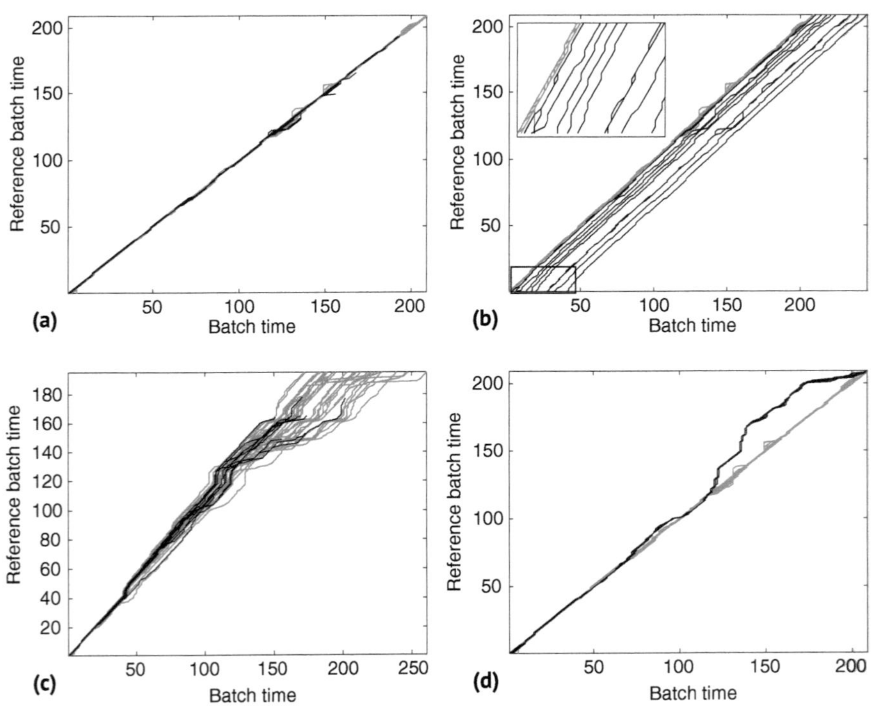

Figure 4.23 Warping information derived from the Multisynchro-based synchronization of the raw batch trajectories for each of the asynchronism scenarios: (a) Case #1: class I (gray lines) and class III (black lines) asynchronisms; (b) Case #2: class I (gray lines) and class IV (black lines) asynchronisms; (c) Case #3: class II (gray lines) and class III (black lines) asynchronisms; and (d) Case #4: class I asynchronism (gray and black lines). The gray lines represent the first dataset of 30 batches, and the black lines represent the second dataset of 10 batches.

Figure 4.23b). Finally, the warping profiles associated with Case #4 are very similar to the warping profiles from DTW at a first glance, although subtle differences are observed at certain phase of the process (see Figure 4.23d).

To illustrate whether there is an effect of the type of synchronization on the synchronized trajectories, the batch data affected by the four cases of asynchronisms are compared with the outcomes of the synchronization based on the DTW and Multisynchro algorithms in Figure 4.24. The first remarkable finding for the NOC datasets simulated in this study is the similarity of the results in cases #2 and #4 (see Figure 4.24d–f and j–l, respectively). Only subtle differences are observed at certain phases of the batch runs. Note, however, that if the batch trajectories are perturbed by an anomaly (e.g., spike or subtle drift) in the shifted part of the trajectory, the DTW algorithm would shrink the corresponding batches at that time interval by averaging the measurements of the J process variables. Most likely, the abnormal feature not belonging to the batch processing might remain in the trajectories, which might produce a false alarm. In contrast, the Multisynchro algorithm would automatically discard the first sampling times, thereby averting the addition of non-process features.

Cases #1 and #3 indicate major differences in the final result due to the presence of batches affected by class III asynchronism (see Figure 4.24a–c and Figure 4.24g–i, respectively). When the batches are incomplete, the DTW algorithm correctly synchronizes the batch trajectories from the initial point (1, 1) to the optimum last matching point (k_{ref}, K_i) (the last closest point of the black lines to the diagonal profile in Figure 4.22c). From the $(k_{\text{ref}} + 1)$th to the K_{ref}th sampling time of the reference batch, the last point of the ith batch is matched, leading to the vertical transitions observed (see Figure 4.22c). This would lead to expansions of the batch trajectories, i.e., the addition of replicated values of the K_ith sampling time in the ith batch. Consequently, flat profiles in the process variables – i.e., replicated values of the last actual value – are introduced (see Figure 4.24b, h). This is an artifact since the batches were not actually finished and the remaining trajectory till completion is computed in an inappropriate way. Furthermore, these inaccuracies are inherited in the synchronization. Note that when a batch is finished earlier than the historical batches, the addition of artifacts in data may be higher. This would cause a possible change of the trajectory profile. In the batch data simulated, the largest cut was approximately 50 sampling times. As observable in Figure 4.24b, h, it changes the shape of the profiles in the second half of the batch runs and, consequently, in the normal process pace.

The higher the addition of artifacts, the higher the uncertainty inherited. This variability may severely affect the interpretation of the subsequent multivariate statistical model, and, therefore, the performance of the monitoring scheme. A detailed study on the effects of the synchronization on bilinear process modeling can be found in the literature.[70]

Several publications in the literature and commercial software packages suggest that only batches with different durations are subject of synchronization. However, when some key process events are not fully aligned, regardless of the batch duration, the batch trajectories must be synchronized. This is clearly shown in Case #4 represented in Figure 4.21d. As evidenced in this figure, the acetate concentration

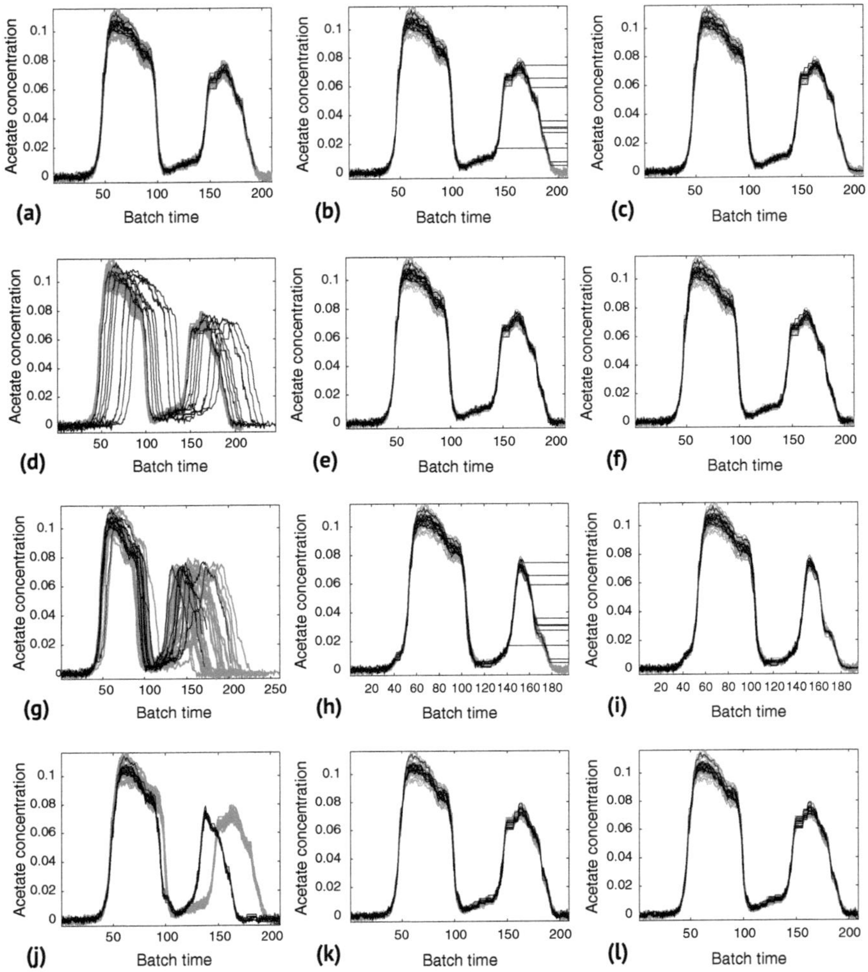

Figure 4.24 Comparison of the acetate concentration of 40 NOC raw batches (a, d, g, and j) with their respective trajectories synchronized by DTW (b, e, h, and k) and by Multisynchro (c, f, i, and l) in four different asynchronism scenarios: (a,b, and c) Case #1: class I (gray lines) and class III (black lines) asynchronisms; (d, e, and f) Case #2: class I (gray lines) and class IV (black lines) asynchronisms; (g, h, and i) Case #3: class II (gray lines) and class III (black lines) asynchronisms; and (j, k, and l) Case #4: class I asynchronism (gray and black lines).

trajectory shows that the second half of the batch run (from the 120th sampling time onward) has a different pace for the two groups of batches, denoted as black and gray lines. From the start to the 90th sampling time, the main process phenomena apparently occur at the same sampling time across batches. To ensure that the key process events are actually synchronized, batch synchronization should be applied to batch data. Figure 4.22d illustrates the resulting warping profiles from the synchronization at the high-level step. As observable, there are two groups of profiles clearly distinguished; those corresponding to the 10 batches where a different process pace was forced (black profiles) and the rest of the batches (gray lines). Looking at the profiles corresponding to the 10 batches with asynchronism, one can observe two main time periods of large deviation from the main diagonal: from the 75th to the 100th sampling time and from the 120th to the end of the batch, the deviation being smaller in the former than in the latter. In both time periods, these warping profiles have a higher number of vertical transitions than horizontal transitions. This is the reason why these warping profiles are beyond the main diagonal. It indicates that the batches with asynchronism had a slower process pace than the rest. Hence, the synchronization algorithm needs to expand the corresponding batch trajectories at the aforementioned batch time periods.

In Case #4, batch synchronization is rarely applied because batches already have the same duration. There is a commercial software for batch process monitoring, e.g., SIMCA Release 13.0.3,[165] which only demands the synchronization of the batch trajectories when they have different lengths[5]. To emphasize the importance of this step, the raw batch trajectories are again compared with those obtained from the batch synchronization based on the Multisynchro algorithm, whose results are very similar to those obtained with DTW. For comparative purposes, two different process variables are illustrated in Figure 4.25. As noticeable, the raw trajectories of the specific oxygen uptake (see Figure 4.25a) and carbon dioxide evolution (see Figure 4.25c) rates belonging to 10 out of the 40 raw batches (black lines) differ from those corresponding to the rest of the batches (gray lines). These differences are mainly shown in the last stage of the process, from the 120th sampling time onward, which suggests that the fermentation in the second half of the process took less time than in the rest of the batch trajectories. Hence, the synchronization of this stage is needed for subsequent analysis. Once the Multisynchro approach is applied to batch data, the resulting 40 profiles not only have equal lengths, but also the segments of the profile corresponding to the last process stage overlap across batches; see the synchronized trajectories of the specific oxygen uptake and carbon dioxide evolution rates in Figure 4.25b, d, respectively.

4.3 Effects of Synchronization on the Correlation Structure

The synchronization of the batch trajectories, along with the equalization of process variables, is the first step in the modeling cycle illustrated in Chapter 1. Incorrect and/or inaccurate synchronization may produce additional artifacts, such as spikes or dumps, artificial shifts, and fictitious correlations among process variables. This inaccuracy may seriously affect the outcome of the subsequent steps in the modeling, and with it, the understanding of the process, the prediction of process variables or properties, and the detection of abnormal situations.

This section is aimed at studying the differences between some of the most used synchronization techniques in the field of process chemometrics, namely IV, TLEC on the whole trajectories (TLEC) and between defined landmarks (TLEC-events),

---

5 The main synchronization procedure used in SIMCA Release 13.0.3 is the so-called time linear expanding/compressing (TLEC)-based method, which is based on linearly expanding and/or compressing pieces of variable trajectories in the local batch time dimension.[46] In case the differences in batch length are greater than 20%, a maturity variable is used as the basis of batch synchronization instead of the local batch time.[46]

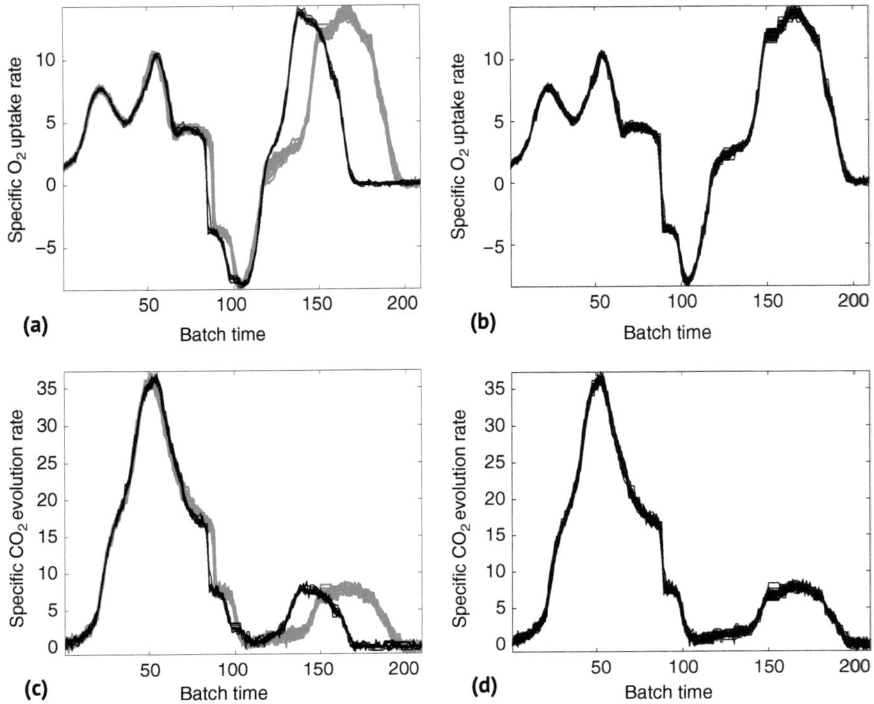

Figure 4.25 Batch trajectories belonging to the process variables specific oxygen uptake rate (a, b) and specific carbon dioxide evolution rate (c, d) without applying any synchronization (a and c, respectively) and applying the Multisynchro approach (b and d, respectively). The black lines in (a) and (c) represent the raw trajectories belonging to 10 out of 40 batches with the Case #4 asynchronism embedded. *Source*: Reproduced from Ref.[66], with permission of John Wiley & Son.

and DTW. In particular, two synchronizations based on two different indicator variables are performed. The first synchronization uses the fermentation rate as the indicator variable, and non-uniform increments are defined. In the second IV-based synchronization, the biomass concentration is used as the indicator variable with constant increments calculated between the start of the batch and the end common value across all batches; see Section 4.2.1 for further details. TLEC-based synchronization is carried out by linearly interpolating all the data points. In regard to TLEC-events, ten key process landmarks are defined based on the evolution of the process variables, and linear interpolation is performed in each defined stage. For DTW-based synchronization, the conditions and constraints are set as defined in Ref.[67, 88] Additionally, a second synchronization is applied to the synchronized batch trajectories by IV and TLEC using DTW (IV-DTW and TLEC-DTW). This re-synchronization is executed to check if the application of nonlinear synchronization techniques that take into account the presence of key process events (the so-called stretching, compressing, and translating (SCT)-based methods) enhances the synchronization results.

To study the effect of the synchronization on the relationship between the process variables, we will focus on the change of the correlation structure. Different metrics can be defined for comparison purposes, such as: the total, dynamic partial, and instantaneous-dynamic partial maps. The former calculates the variance-covariance matrix of the synchronized, preprocessed, and batch-wise unfolded two-way array that contains the batch trajectories. This matrix provides practitioners with a comprehensive view of the dynamic relationships in the batch data, including not only the instantaneous variances and cross-variances of the variables at each sampling time, but also the auto-covariances and lagged cross-covariances. Revealing the time-varying dynamics helps practitioners better understand the relationships between process variables over time. The dynamic partial variance-covariance map is the computational result of the partial covariance for all the possible lags in time. It is useful to observe dynamic relationships without taking into account the instantaneous relationships between process variables. This information should be used when the objective of the modeling is to predict the current value of a variable from past measurements of the process (e.g., to perform one-step-ahead predictions). Finally, the instantaneous-dynamic partial variance-covariance map contains the partial covariance considering the instantaneous relationships of the process variables. The resulting map is useful when the goal is to obtain parsimonious models for process monitoring that properly capture the complex dynamics by adding the optimal number of lags. The code for the calculation of this matrix is also available in the MVBatch toolbox.[65]

The comparison of the synchronization methods is carried out by taking as a reference the optimal synchronization for this dataset. The optimal synchronization of the batch trajectories consists of shrinking, compressing, and nonlinearly translating segments of data, similarly to the DTW synchronization, in accordance with the evolution of the simulation time. Recall that the latter is the original time of processing warped in the simulation to make all batches different in length. The total variance-covariance matrix $\mathbf{S}_j$ of each dataset synchronized by the selected methods is then estimated for the ten process variables. The dissimilarity between each of these variance-covariance matrices $\mathbf{S}_j$ and the one calculated from the optimum synchronization $\mathbf{S}_j^*$ is assessed by calculating the correlation matrix distance (CMD) index[78] as follows:

$$d(\mathbf{S}_j, \mathbf{S}_j^*) = 1 - \frac{\mathrm{Tr}(\mathbf{S}_j \cdot \mathbf{S}_j^*)}{\| \mathbf{S}_j \|_F \cdot \| \mathbf{S}_j^* \|_F} \tag{4.18}$$

where $\mathrm{Tr}(\cdot)$ represents the trace of a $(K \times K)$ matrix, $\| \cdot \|_F$ is the Frobenius norm and K the number of sampling times. This dissimilarity index varies from zero to one. It becomes zero if the variance-covariance matrices are equal to a scaling factor, and one if they differ to a maximum extent. Hence, the CMD index is a good indicator of how well a specific synchronization technique is able to align the features of the trajectories, minimizing the addition of artifacts, and therefore, reducing the changes in the original correlation structure.

The total variance-covariance matrices of the batch data synchronized by IV using the fermentation rate as the indicator variable, and by IV-DTW corresponding to process variables such as glucose concentration, acetate concentration, and specific oxygen uptake rate are shown in Figure 4.26. At first glance, there are differences in the relationships among the variables over time after the two synchronization approaches were applied, which are more significant in the last two process variables (see Figure 4.26b, e and c, f for comparison). In the case of the glucose concentration, a positive correlation is observed (see the red square located on the main diagonal at the first stage of the process – till the 42nd sampling time approximately – in Figure 4.26a, d). The variance-covariance matrix of the glucose concentration synchronized by IV shows a change in the process dynamics from the 42nd to the 54th sampling time. In particular, positive lagged cross-covariances of low order are observed (see the fuzzy red diagonals outside the main diagonal from the 42nd to the 54th sampling time in Figure 4.26a). Based on these observations, an apparent change occurs within the above-mentioned interval. However, the results of IV-DTW reflect a gradual diminution of the correlation in the first phase, which is more an indication of a transition between phases, rather than a complete change of phase (see the fuzzy red area covering the red square in Figure 4.26d. These differences might be caused by partially inaccurate transformation of the timescale into the IV scale using interpolation. To verify this claim, the synchronized trajectories of the three process variables with their respective average trajectories are shown in Figure 4.27. As appreciated from

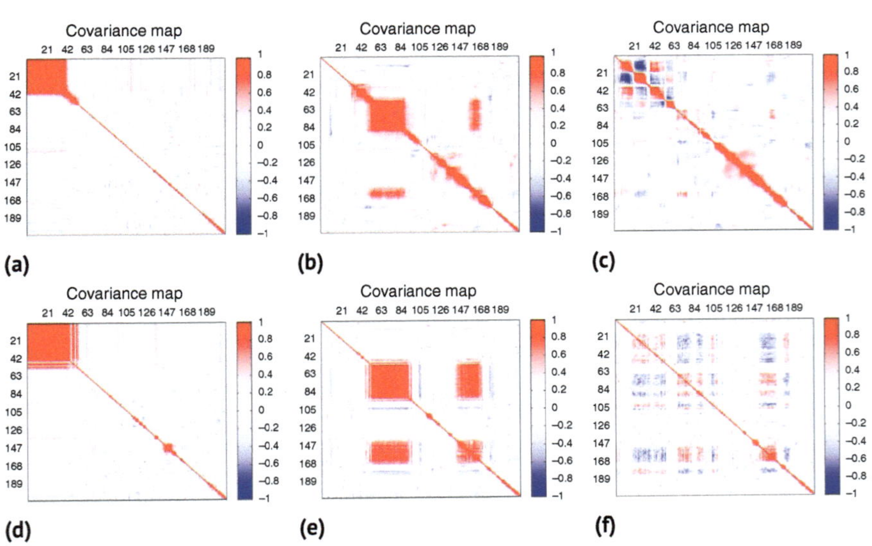

Figure 4.26 Total variance-covariance matrices calculated for the glucose concentration (a,d), acetate concentration (b,e), and specific oxygen uptake rate (c,f). These process variables were previously preprocessed with trajectory C&S[6] and batch-wise unfolded[7] after synchronizing with IV using the fermentation rate as the indicator variable (a,b,c) and a second synchronization using DTW (IV-DTW) (d,e,f).

6 see Chapter 5 for explanation in this preprocessing method.
7 see Chapter 6 for details in the unfolding method.

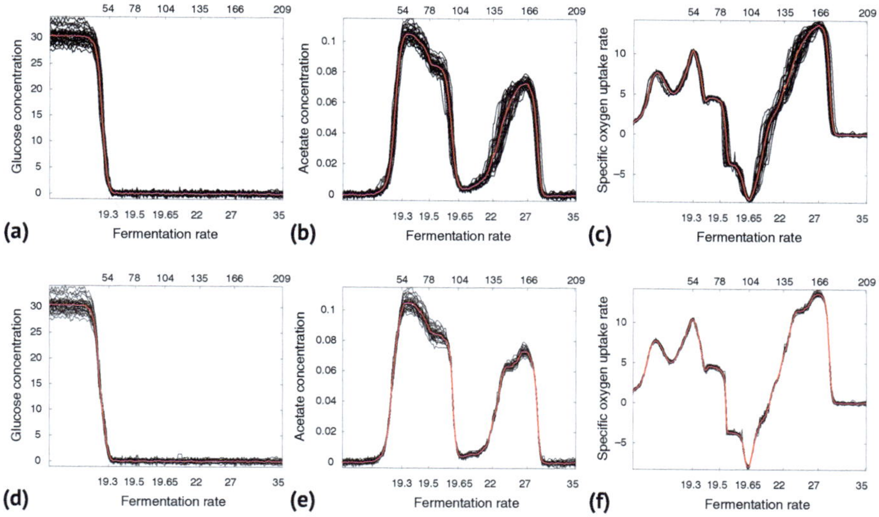

Figure 4.27 Batch trajectories of the glucose concentration (a,d), acetate concentration (b,e), and specific oxygen uptake rate (c,f) synchronized by IV (a,b,c) and IV-DTW (d,e,f), expressed as a function of the IV evolution (lower x-axis) and the number of interpolated samples (upper x-axis). Red lines denote the average trajectory for each process variable.

Figure 4.27a after IV synchronization, the transition between the excess in glucose (maximum values) and the depletion of this substrate (minimum values) – i.e., from the 42nd to the 54th sampling time – shows some variability around the average trajectory. If these trajectories are subsequently aligned by using nonlinear synchronization steps to overlap the landmarks (IV-DTW synchronization), the variability is slightly reduced (see Figure 4.27d). As observed in this example, the increase of variability due to synchronization might affect the correlation data structure. Consequently, it might jeopardize the segregation of the phases and the interpretation of the multivariate models for process understanding, although it might not necessarily impact the detection of anomalies.

In the glucose concentration, we have seen slight changes in the process dynamics. However, one of the major effects of misalignment is the creation of nonexistent relationships over time. Upon interpreting the variance-covariance matrices estimated from the trajectories synchronized by IV and IV-DTW (see Figure 4.26b, c, e, and f, respectively), some differences are found. In the acetate concentration for IV-based synchronization, there seem to be two main phases; the first phase ranges from the 32nd to the 105th sampling time approximately, and the second phase ranges from the 116th to the 175th sampling time approximately (see the fuzzy red square at the top-left side and the square at the bottom-right side with positive correlation of diverse magnitude in Figure 4.26b). Between these intervals there seems to be a transition due to a mild negative correlation. However, the fuzziness of these intervals on the variance-covariance map, jointly with the changing correlations over the batch evolution, leads us to consider the existence of diverse phases. In the first interval, strong positive instantaneous variances and cross-covariances at every sampling time point (see dark red color in the main diagonal), as well as strong positive auto-covariances and lagged cross-covariances of diverse order (see the red squares located on the main diagonal), are observed. In particular, three main phases of different duration appear to exist (see the red squares at subintervals [32, 54], [55, 90], [91, 105] in Figure 4.26b). In the second interval, there are strong positive instantaneous variances and cross-covariances at the subintervals [116, 130] and [131, 151], with dynamics of low order (see the red area at the bottom-right side outside the main diagonal in Figure 4.26b). These correlations can be explained and confirmed by looking at the synchronized trajectories depicted in Figure 4.27b. As can be observed in this figure, the two main intervals coincide with the process stages at which acetate is excreted (see intervals [32, 105] and [116, 175] in Figure 4.27b). However, the change of variability within these intervals produced by synchronization is the main reason the dynamics vary over time. Even though IV aligns the key process events across batches, the inaccuracy of interpolation does not allow for a clear revelation of the actual correlation structure. Improving the results of the IV-based synchronization with a DTW-based synchronization, the two phases previously pinpointed are better reflected in the variance-covariance matrix (see the large red squares on the main diagonal in Figure 4.26e). Note that a negative correlation is unmasked at the [100, 104] and [175, 180] intervals, which seems to be the transition among phases, as appreciated from Figure 4.27e. Thanks to a notable reduction of the variability produced by the IV synchronization, the actual relationships between the process variables are captured.

Another type of change in correlation is found in the specific oxygen uptake rate. At the start of the batch, IV aligns properly the events driving the process, although it increases the variability around the average trajectory (see Figure 4.27c). As a consequence, the resulting variance-covariance matrix of the IV-synchronized data shows fictitious correlations over the evolution of the batch. First, alternating positive and negative correlations from the start of the batch until the 63rd sampling time are observed (see the top-left area in Figure 4.26c). These time-varying correlations are mainly due to alternating intervals at which the batch trajectories evolve above and below the average trajectory (see first 63 sampling times in Figure 4.27c). In addition, two main phases seem to be outlined, the first phase ranging from the 64th to the 105th sampling time, and the second phase from the 106th to the 180th sampling time (see the positive correlation depicted as red squares at the aforementioned intervals lying on the main diagonal in Figure 4.26c). These intervals are associated with the process stages in which the variability around the average trajectory is higher (see Figure 4.27c). After a re-synchronization, this variability is notably reduced, thereby revealing the actual correlations over the batch run for the specific oxygen uptake rate (see Figure 4.26f and 4.27f).

In Section 4.2.1, the effects of an inappropriate selection of the indicator variable and of the IV levels at which data are interpolated were shown on the resulting trajectories. To study the impact on the correlation structure, the total variance-covariance matrices of the batch data synchronized by the IV method with the biomass concentration as the indicator variable, and by the IV-DTW method are provided in Figure 4.28. Comparing these total variance-covariance matrices with those estimated from the data synchronized with the fermentation rate as the indicator variable, there are significant differences. As expected, when the shape of the batch

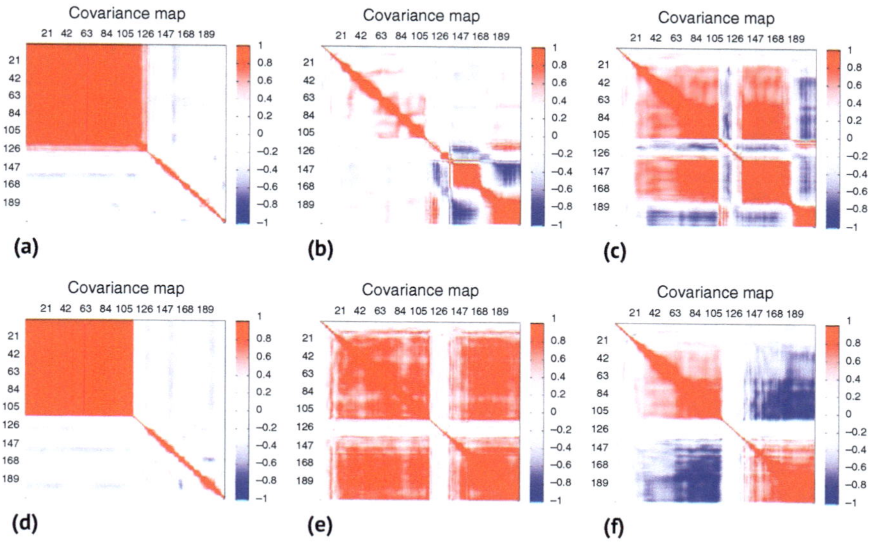

Figure 4.28 Total variance-covariance matrices calculated for the glucose concentration (a,d), acetate concentration (b,e), and specific oxygen uptake rate (c,f). These process variables were previously preprocessed with trajectory C&S and batch-wise unfolded after synchronizating with IV using the biomass concentration as the indicator variable (a,b,c) and after a second synchronization using DTW (IV-DTW) (d,e,f).

trajectories is modified, the actual correlation structure is completely perturbed. The relationships found in the synchronized data are not anymore representative of the phenomena occurring in the process. However, for illustrative purposes, it is worth studying the differences found between the synchronizations based on IV and IV-DTW.

In the case of the glucose concentration, a positive correlation is observed over time (see the red rectangles located on the main diagonal at the first stage of the process – till the 110th sampling time point approximately – in Figure 4.28a, d). Comparing the two resulting matrices from the two different synchronizations, it can be observed that the duration of these correlations varies. In particular, the duration is longer in those correlations calculated from data synchronized by IV than by IV-DTW. This difference is due to inaccuracies in the synchronization caused by the interpolation conducted at equal-distanced intervals, which not only changes the variable trajectories but also increases variability around the average trajectory. To support this claim, the synchronized trajectories of the three process variables with their respective average trajectories are shown in Figure 4.29. As can be appreciated from Figure 4.29a after IV synchronization, the sampling time at which the glucose concentration reaches its minimum value (end of the phase) considerably varies from batch to batch (from the 110th to the 130th sampling time). If these trajectories are subsequently synchronized, taking into account the overlap of the landmarks (IV-DTW synchronization), the variability is reduced (see Figure 4.29d).

In the acetate concentration, apparently there are two phases, the first phase ranging from the start of the batch until the 125th sampling time approximately and the second phase from the latter sampling time onward (see the fuzzy red square at the

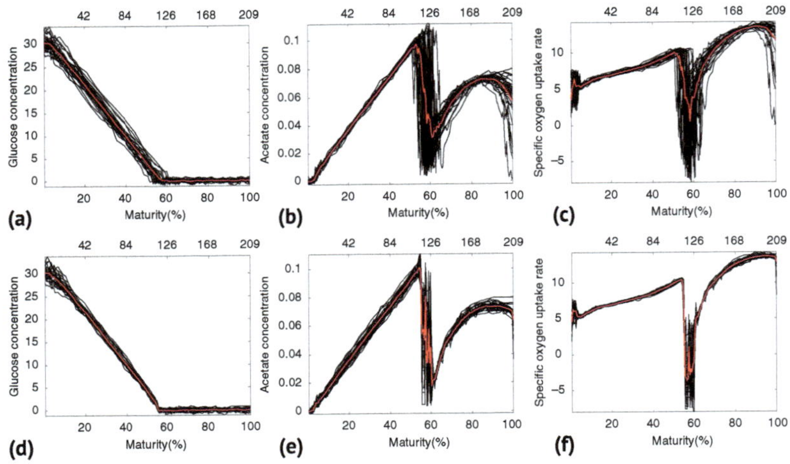

Figure 4.29 Batch trajectories of the glucose concentration (a,d), acetate concentration (b,e), and specific oxygen uptake rate (c,f) synchronized by IV (a,b,c) and IV-DTW (d,e,f), expressed as a function of the process maturity (lower x-axis) and the number of interpolated samples (upper x-axis). Red lines denote the average trajectory for each process variable.

top-left side and the square at the bottom-right side formed by red and blue sub-squares in Figure 4.28b). In the first phase, strong positive instantaneous variances and cross-covariances at every sampling time (see the dark red color in the main diagonal), as well as strong positive auto-covariances and lagged cross-covariances of low order (see the red rectangles located outside the main diagonal), are observed. The latter becomes milder over time, as appreciated from the red fuzzy rectangles outside the main diagonal. In the second phase, there is a strong correlation in the intervals [140, 165] and [180, 209], which are negatively correlated to each other (see blue rectangles at the bottom-right side outside the main diagonal in Figure 4.28b). Between the phases there seems to be a transition due to the lack of correlation in the interval [111, 130]. These correlations can be explained and confirmed by examining the synchronized trajectories depicted in Figure 4.29b. As can be observed in this figure, the trajectories evolve over time following the average trajectory till the 100th sampling time. At this point, the key process events are not overlapped across batches, since all the trajectories are shifted. This misalignment is also observed in the next landmark located around the 110th sampling time. For thirty sampling times, there is apparently no clear relationship over time due to an inaccurate synchronization, thereby producing a transition among phases. In the second stage, from the 140th to the 165th sampling time, most of the batch trajectories are beyond the average trajectory, whereas from the 180th to the 209th sampling time, it appears that most of the trajectories remain below the average. This observation explains the negative correlation found among intervals at the second phase in the variance-covariance matrix in Figure 4.28b. Improving the results of the IV-based synchronization with a second DTW-based synchronization, the two phases previously pinpointed are better reflected in the variance-covariance matrix (see the large red squares on the main diagonal in Figure 4.28e). However, in contrast to the IV-based synchronization, the correlation of the variable is positive at the second stage. Thanks to a notable enhancement in synchronization, the actual relationships become apparent.

Let us study the implications of using TLEC-based methods to synchronize batch trajectories. Figure 4.30 provides the variance-covariance matrices of the three process variables trajectories synchronized by TLEC, TLEC-DTW, and TLEC-events. Comparing the matrices, we find significant differences between those synchronized by the former (see Figure 4.30a–c) and the rest (see Figure 4.30d–i), where in the acetate concentration and specific oxygen uptake rate, the differences are more significant. As can be appreciated from Figure 4.30b, c, there are positive instantaneous variances and cross-covariances at every sampling time (see main diagonal), but alternating positive and negative correlations over time (see alternating blue and red areas along the diagonal outside the main diagonal). This is a clear symptom of, first, a lack of synchronization of the landmarks, and second, an excessive number of nonlinear shifts in the trajectories that mainly produce this type of relationship

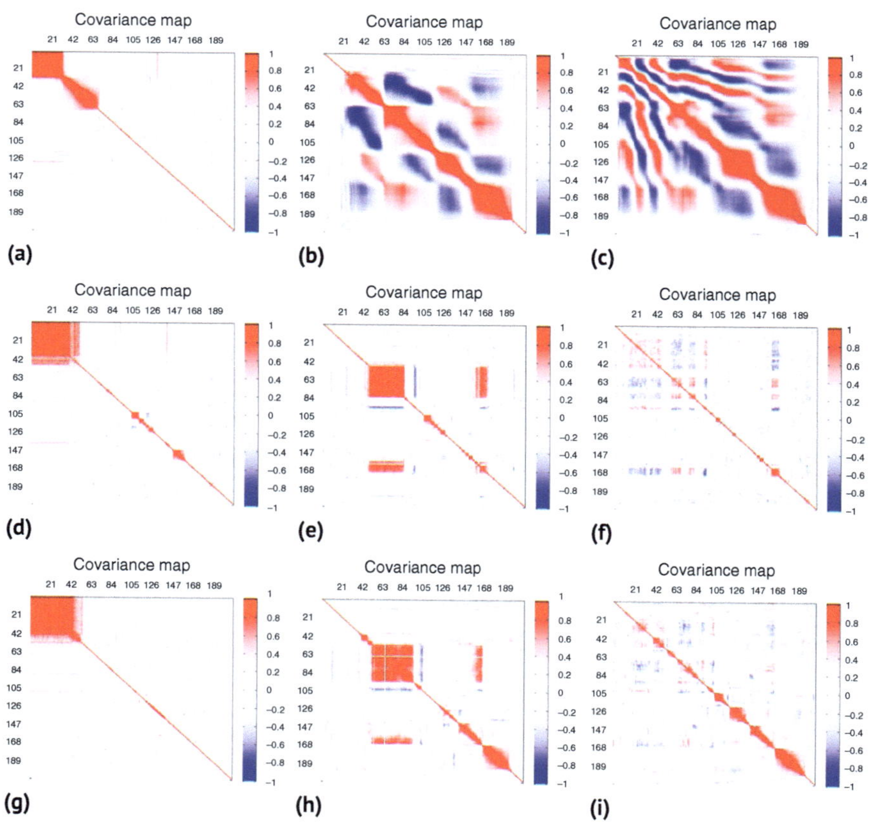

Figure 4.30 Total variance-covariance matrices calculated for the glucose concentration (a,d,g), acetate concentration (b,e,h), and specific oxygen uptake rate (c,f,i). These process variables were previously preprocessed with trajectory C&S and batch-wise unfolded after synchronizing with TLEC (a,b,c), TLEC based on aligning predefined events (TLEC-events) (d,e,f), and TLEC with a subsequent second synchronization using DTW (TLEC-DTW) (g,h,i).

among variables and lagged measurements. For confirmation, the respective trajectories synchronized by TLEC, are plotted in Figure 4.31. Apart from showing a set of asynchronous trajectories, the nonlinear shift produced by the synchronization is greater in the acetate concentration (see Figure 4.31b) than in the specific oxygen uptake rate (see Figure 4.31c). When a second synchronization is performed on the trajectories already synchronized by TLEC, the quality of the synchronization improves (see Figure 4.30e, f for TLEC-events and Figure 4.30h, i for TLEC-DTW). The uncommon correlation over time is completely diluted, revealing the actual relationships (see Figure 4.31e, f for TLEC-events, and Figure 4.31h, i for TLEC-DTW).

The last synchronization method applied to the dataset under study is the DTW. The resulting variance-covariance matrices for the three selected variables are plotted in Figure 4.32. As can be appreciated by comparing Figure 4.26 and 4.30 with Figure 4.32, the results observed for IV-DTW, TLEC-DTW, and TLEC-event are similar to those obtained for DTW to a certain extent. However, IV-DTW seems to better approximate the variance-covariance matrix of the data synchronized by DTW. The reason for this enhancement is that the re-synchronization by SCT-based methods

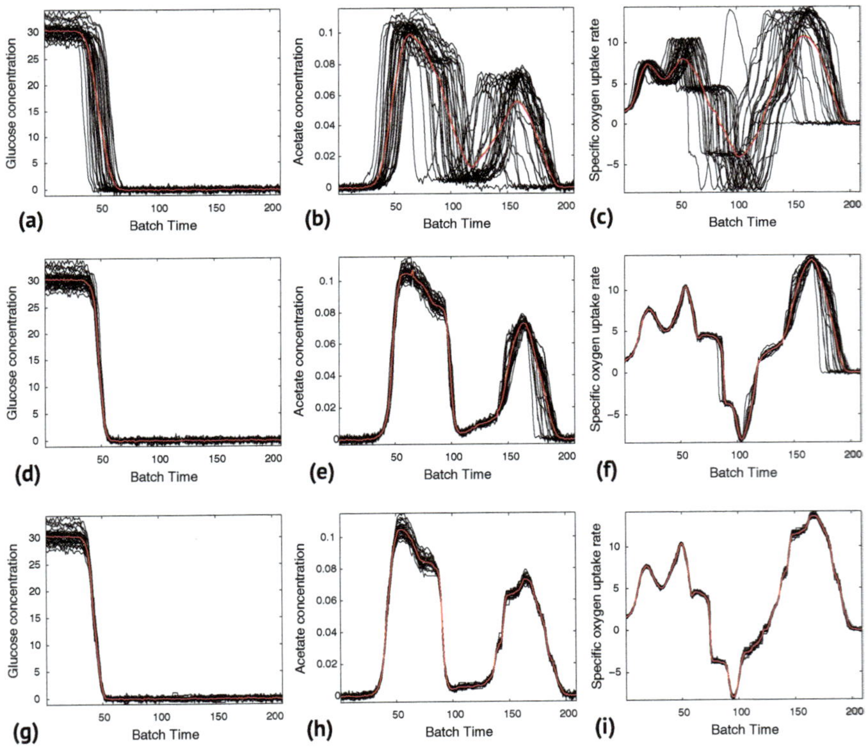

Figure 4.31 Synchronized batch trajectories of the glucose concentration (a,d,g), acetate concentration (b,e,h), and specific oxygen uptake rate (c,f,i) by TLEC (a,b,c), TLEC-events (d,e,f) and TLEC-DTW (g,h,i). Red lines indicate the average trajectory for each process variable.

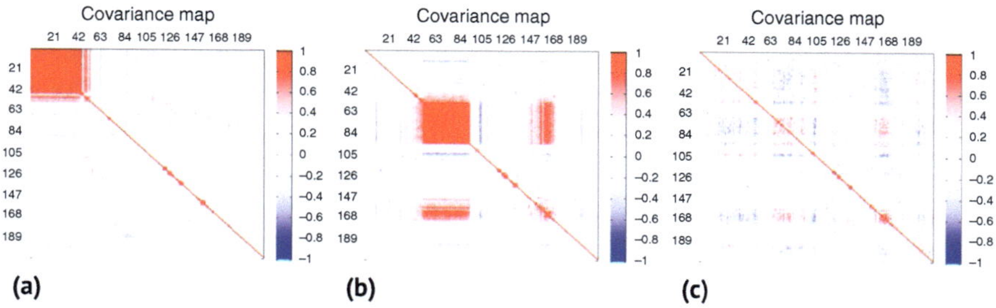

Figure 4.32 Total variance-covariance matrices calculated after synchronization using the DTW algorithm, trajectory C&S preprocessing and batch-wise unfolding three process variables: (a) glucose concentration, (b) acetate concentration, and (c) specific oxygen uptake rate.

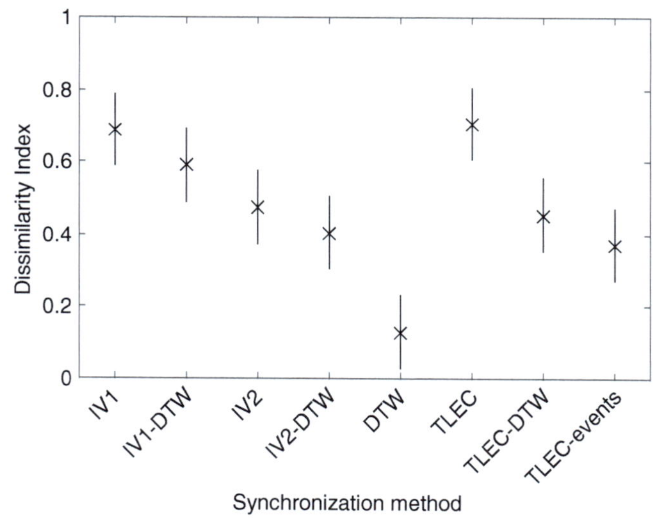

Figure 4.33 LSD intervals for the dissimilarity index among the variance-covariance matrices estimated for each process variable using the optimum synchronization and the synchronization based on IV, IV-DTW, DTW, TLEC, TLEC-DTW, and TLEC-events.

of nonoptimal synchronized trajectories reduces the variability produced by IV synchronization, and ensures the overlap of key process events in the TLEC synchronization. Hence, the actual relationship of each variable over the evolution of the batch is significantly better captured.

With the aim of studying whether there are statistically significant differences between the synchronization approaches under study in terms of changes in the correlation structure, an ANOVA is carried out. In this case, the variance-covariance matrices of all process variables are calculated for IV and IV-DTW using as indicator variable the biomass concentration (IV1 and IV1-DTW), and the fermentation rate (IV2 and IV2-DTW), DTW, TLEC, TLEC-events, and TLEC-DTW. The dissimilarity index between the variance-covariance matrices of each process variable for each synchronization method and the variance-covariance matrix obtained from data obtained through the optimal synchronization, as explained in Section 4.3, is used as the response variable in ANOVA. The study shows statistical differences in the changes of the correlation structure based on the synchronization method employed (p-value < 0.05). The LSD intervals are shown in Figure 4.33. DTW synchronization clearly yielded the best results since the associated dissimilarity index is close to 0 and there are statistically significant differences with the rest of the methods. In contrast, no statistical differences are found between IV and IV-DTW, and between TLEC-DTW and TLEC-events (although IV-DTW and TLEC-DTW seem to provide slightly better results than IV and TLEC, respectively). However, these methods clearly outperform the results from the classical TLEC method in terms of changes in the correlation structure. Hence, the better the synchronization of the landmarks, the less change in the correlation structure.

5

Batch Data Preprocessing

After batch data alignment (Chapters 3 and 4), and prior to data modeling (Chapter 6) and monitoring (Chapter 7), data often needs to be preprocessed to ensure that model and monitoring statistics are focused on the variance of interest. In the literature, there has been extensive discussion on the very nature of this variance, and which preprocessing operation is the most favorable for estimating it. In the following sections, we discuss several of these operations and their effects on batch data.

There are a lot of classes of sensors in the industry. Each of them has its own common preprocessing methods to prepare data for subsequent analysis. A typical case is when dealing with spectral data,[147] where signal correction methods are of paramount importance. This section is not intended to revise the different preprocessing methods for each type of sensor. Rather, the aim is to explain the implications of the typical preprocessing operations in engineering-type process variables such as temperatures, pressures, etc. A relevant discussion on this topic can be found elsewhere.[21, 40, 180]

5.1 Batch Preprocessing Operations

Depending on the nature of batch data and the type of model to be fitted, the preprocessing approach may be different.[72] As already mentioned in Chapter 1, two main preprocessing methods are widely used in process chemometrics: trajectory-wise centering and scaling (trajectory C&S) and variable-wise centering and scaling (variable C&S). Both operations perform mean centering and scaling to unit variance, which is often referred to as auto-scaling, but they differ in the data block considered, see Figure 5.1.

Trajectory C&S consists of auto-scaling the data corresponding to each vector $\mathbf{x}_{jk}$, containing the measurements of the jth process variable at the kth sampling point for all the batches, Figure 5.1a. Provided the aligned three-way data array contains a total of J variables, K sampling times, and I batches, JK averages and standard deviations (for a total of $2JK$ parameters) are estimated from I batches.

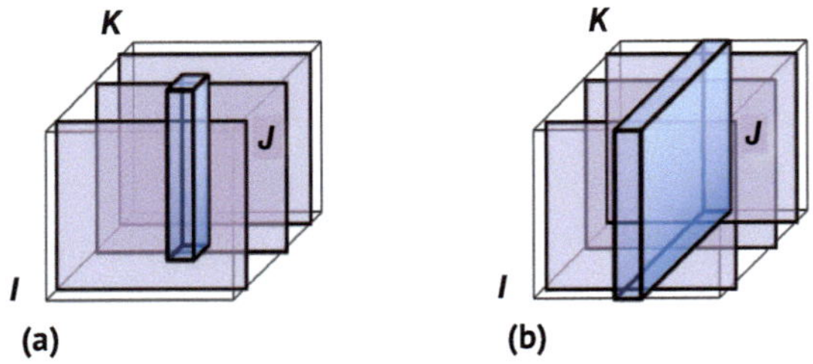

Figure 5.1 Illustration of batch preprocessing operations. Statistics (averages and standard deviations) are computed for each block like the one represented in blue: trajectory-wise centering and scaling (a) and variable-wise centering and scaling (b).

This normalization allows each process variable at each time to have the same weight in the multivariate analysis.

Variable C&S auto-scales the block of data corresponding to each jth process variable. This preprocessing means that each lateral slab $\mathbf{X}_j$ is auto-scaled, as illustrated in Figure 5.1b. In this case, J averages and standard deviations ($2J$ parameters) are estimated from IK each observations. With this normalization, time periods with more variability will be weighted more than periods with less variability (e.g., under tight control) in the multivariate analysis.

The discussion about which of the two preprocessing operations is more adequate has been present in the literature[73, 93, 180] since the two main pioneer research studies in batch multivariate statistical process control (BMSPC)[119, 186] focused on each of them. Nomikos and MacGregor[119] performed trajectory C&S, whereas Wold et al.[186] used variable C&S as a first step of their scheme.

After the desired variability is kept on data, the transformation of the three-way data array to a two-way data array can be carried out as described in the next chapter. For some time, the three-to-two-way transformation and the preprocessing have been understood as a joint operation, so that the preprocessing seemed to be restricted to the auto-scaling of the transformed two-way array. That is a misconception, and the preprocessing methods do not necessarily have to be determined by the transformation to two-way data, although it is important to understand how the combination of these two operations affects the resulting latent variable model (LVM) model.

We see two main advantages of trajectory C&S over variable C&S, which make the former more suitable than the latter for BMSPC: (i) trajectory C&S models the variability around the average trajectory, which is actually the type of variability of interest to monitor a batch process[93]; and (ii) the model is simplified since a batch process is nonstationary along the time axis, but it is expected to be quasi-stationary[1] along the batch mode, which is sometimes referred to as cyclostationary. By removing the average trajectory and normalizing the variance around it (i.e., by applying trajectory C&S), we transform the batch data from its original nonstationary form to a quasi-stationary form, which is optimal for the future detection of anomalies and simplification of the monitoring system design.

1 Provided the batch process is under tight control.

5.2 Mean Centering

As in Section 2.2.1 for bilinear data, we will further discuss the effect of the mean centering operation on the batch data, as well as the effect of the scaling operation.

As previously stated, trajectory centering performs a mean centering of the data corresponding to each process variable (j) at each sampling point (k), while variable centering performs a mean centering of the data corresponding to each process variable (j). From a statistical perspective, the ratio of parameter estimates (based on the number of averages computed) to observations is more favorable in variable centering than in trajectory centering. A more favorable ratio has the beneficial consequence of reducing the variance of the estimates. Nonetheless, the grand mean obtained after variable centering may not be representative of batch data, in particular for nonstationary data along the time mode. Stated otherwise, the grand mean is a biased estimate. In particular, a good trade-off between bias and variance may be obtained by considering more sophisticated preprocessing operations, consistent with some of the three-to-two-way transformations discussed in the next chapter.

Both mean centering operations (trajectory-wise and variable-wise) are illustrated in Figure 5.2 for an artificial dataset representing a single variable in a batch process.

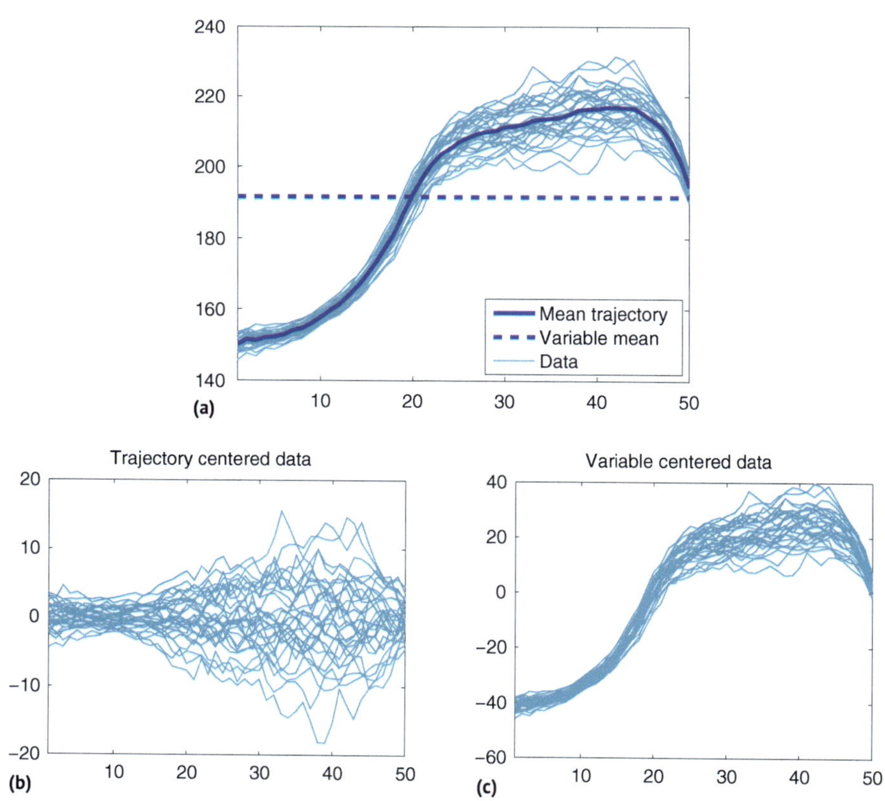

Figure 5.2 Effect of centering in a single variable of a batch dataset (a) for trajectory-wise centering (b) and variable-wise centering (c).

Figure 5.2a shows the original, raw data, with a nonstationary profile in the time mode. Figure 5.2b shows that trajectory centering removes the mean trajectory, and only the heteroscedasticity in the variance along the time mode remains. This operation is a requirement when one would like to detect a deviation from the golden standard trajectory by inspecting the magnitude of the preprocessed data (e.g., using typical multivariate statistics). Figure 5.2c illustrates that, after variable centering, the mean trajectory is centered at 0, but its shape remains in the data. Clearly, with this operation, we cannot directly assess the deviation from the golden standard trajectory using the magnitude of the preprocessed data. For instance, at time point 40, a batch following the exact average trajectory preprocessed by trajectory centering (Figure 5.2b) would have a magnitude of 0, while the same batch preprocessed by variable centering (Figure 5.2c) would be separated from the origin of coordinates in 20 units.

5.3 Scaling

Auto-scaling is commonly used when there is no a priori knowledge of the relative importance of the different pieces of batch data, like relevant variables or batch phases. Certainly, a different scaling method may be applied if some parts of the process are more important, and therefore are desired to have more leverage in the LVM. On the other hand, the scaling operation is often performed by dividing the data into the same blocks which were used in the mean centering. That is the case in both trajectory C&S and variable C&S. An alternative choice[169] is to combine trajectory centering with variable scaling.

The three approaches are illustrated in Figure 5.3. Figure 5.3a shows that trajectory C&S removes the average trajectory and the heteroscedasticity in the time mode. The resulting data looks like random white noise. Yet, correlations along the time and variable modes remain in this data, while the advantage is that we can easily derive constant control limits, which facilitates the design of monitoring systems. Variable C&S in Figure 5.3b neither removes the mean trajectory, nor the heteroscedasticity. Variable scaling normalizes each process variable j so that it has the same weight in the LVM, but those parts of more variance (e.g., around sampling time 40 in the example) will have a higher leverage on the LVM than parts of lesser variance (e.g., around sampling time 10). Finally, variable scaling after trajectory centering in Figure 5.3c does not remove the time-mode heteroscedasticity, and as such makes the definition of control limits more complicated than in trajectory C&S.

In principle, if nothing is known a priori about the data, trajectory C&S seems to be the best choice for batch data preprocessing. This is because all the periods of the batch evolution and all variables are treated equally. If variable scaling is used, parts of high variance will be the most relevant in the analysis. In principle, the most important information does not have to be of the highest variability. A priori, we want our monitoring system to be equally effective in fault detection during the whole process of evolution. For this, it is beneficial to capture the structure of the data throughout the entire processing of a batch, including parts of low variability. Trajectory C&S seems to be the more appropriate preprocessing for this purpose.

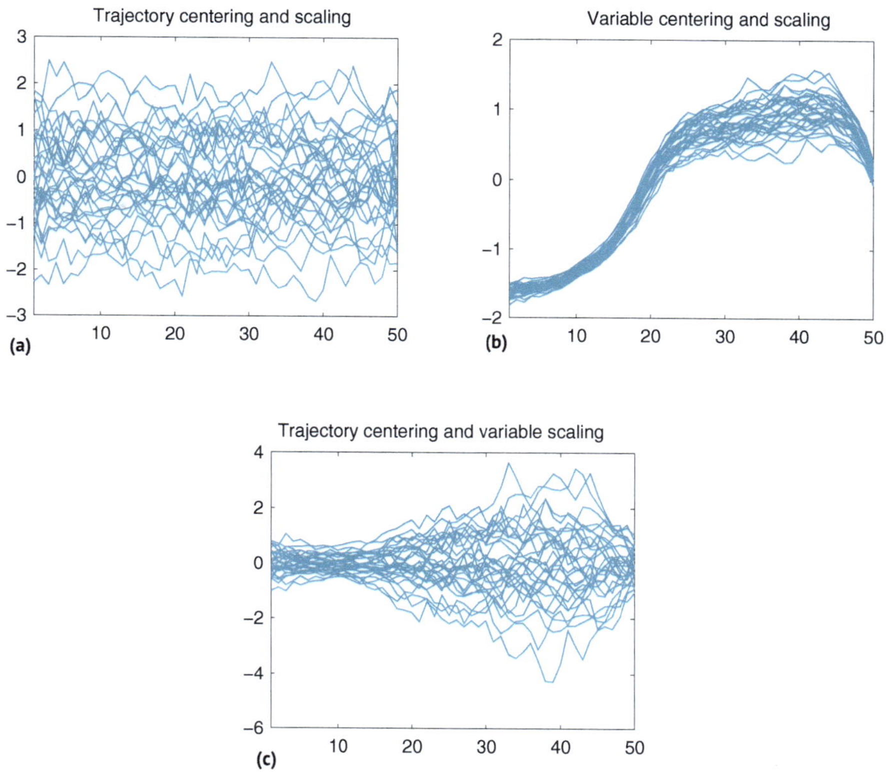

Figure 5.3 Different combinations of mean centering and scaling to unit variance for the example in Figure 5.2: (a) trajectory C&S, (b) variable C&S, and (c) variable scaling after trajectory centering.

Example 5.1: Data preprocessing

This example can be reproduced with the MATLAB® live script called "ex5_1.mlx." As starting point, let us go back to the NOC batch data synchronized with the relaxed greedy time warping (RGTW) algorithm in Chapter 4. Now, different preprocessing methods can be performed on the data. If data contain some missing values, the adequate routine is ||`preprocess3D_nan`||. Otherwise, ||`preprocess3D`|| can be used. Both routines have the same parameters and perform exactly the same preprocessing when data are complete.

```
[cn_alg_p,m,sd] = preprocess3D(cn_alg,prep);
```

The first parameter contains the aligned and unpreprocessed data, and the second parameter identifies the preprocessing method to perform. Possible choices are listed in Table 5.1. The outputs include the preprocessed data, the average estimates and the standard deviation estimates.

Figure 5.4 shows the first three variables of the preprocessed data for trajectory C&S, variable C&S, and trajectory centering and variable scaling. Each batch is represented by a different line along approximately 200 time points, which correspond to the aligned batch duration. We can see that after trajectory C&S, data is auto-scaled along time; after variable C&S, the average trajectory of each variable remains

Table 5.1 Possible choices for **prep** in ||**preprocess3D_nan**|| or ||**preprocess3D**||.

prep	Preprocessing
0	No preprocessing
1	Trajectory centering
2	Trajectory centering and trajectory scaling
3	Trajectory centering and variable scaling
4	Variable centering
5	Variable centering and variable scaling

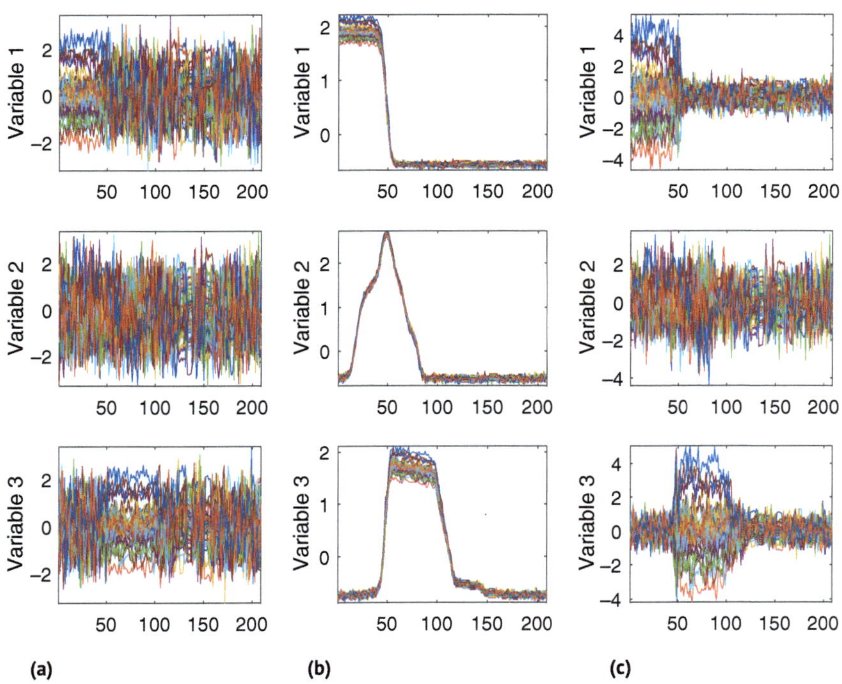

Figure 5.4 First three variables (in the rows) of the data of the *Saccharomyces cerevisiae* fermentation synchronized with the RGTW algorithm and preprocessed following different methods (in the columns). (a) Trajectory C&S (TC&S), (b) Variable C&S (VC&S), and (c) Trajectory centering & variable scaling (TC&VS).

in the data; and after trajectory centering and variable scaling, the preprocessed data is mean-centered along time, but not auto-scaled. Figure 5.5 shows the estimates of the average +/− one standard deviation for the same preprocessing choices. Trajectory C&S captures the average trajectory in the average estimates, and the heteroscedasticity in the standard deviation estimates. Variable C&S only captures the grand mean and grand standard deviation. Trajectory centering and variable scaling capture the average trajectory and the grand standard deviation around it.

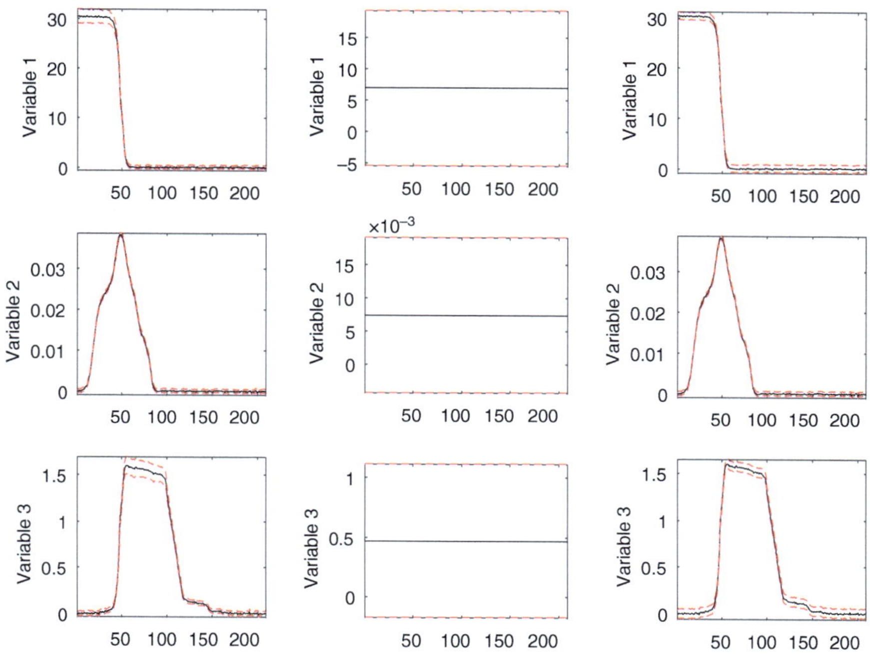

Figure 5.5 Average +/− one standard deviation of the first three variables (in the rows) of the data of the *Saccharomyces cerevisiae* fermentation synchronized with the RGTW algorithm and preprocessed following different methods (in the columns). (a) TC&S, (b) VC&S, (c) TC&VS.

6

Three-way to Two-way Transformation

6.1 Introduction

In model calibration, the aligned and preprocessed three-way data array $\underline{\mathbf{X}}$ $(I \times J \times K)$, which contains the data collected from a batch process at K sampling points on J process variables in I batches, must be rearranged in a number of two-way arrays to fit bilinear latent variable models (LVMs), such as principal component analysis (PCA) or partial least squares (PLS). As commented in Chapter 1, the different approaches to perform this transformation can be classified into three categories: single-model, K-models, and hierarchical model. Most of these approaches were originally defined for PCA-based multivariate statistical process control (MSPC) of batch processes, but they have been extended to regression with PLS.[29]

In this chapter, the theoretical differences among the most important three-way to two-way transformations are illustrated. For this purpose, the covariance matrix of the unfolded two-way data array, i.e., $\mathbf{C}_X = \frac{1}{I-1}\mathbf{X}^T \cdot \mathbf{X}$, is analyzed. As explained in Chapter 2, this matrix is strongly related to the PCA model. In several parts of this book, a LVM has been presented to capture the structure in data, i.e., the relationships among variables. These relationships are extracted from $\mathbf{C}_X$. Depending on the three-way to two-way transformation, $\mathbf{C}_X$ has a different set of variables, which are considered in the analysis. This transformation affects the insights derived from analyses as well as the performance of the monitoring systems and predictability of the resulting LVM. Therefore, it is crucial to fully understand the three-way to two-way transformations to obtain a reasonable model for the data.

The analysis of $\mathbf{C}_X$ is useful to understand the dynamics built in PCA and PCR models. In the case of PLS, it holds that the model is computed from both $\mathbf{C}_X$ and $\mathbf{C}_{Y^TX} = \mathbf{L}^T \cdot \mathbf{L}$, for $\mathbf{L} = \mathbf{Y}^T \cdot \mathbf{X}$. The conclusions presented in this chapter for $\mathbf{C}_X$ are valid for $\mathbf{C}_{Y^TX}$ as well as for PLS regression. The only difference is that in the latter, the relationships are not among the original process variables, but among the covariance values between process and quality variables.

6.2 Single-model Approach

In the single-model approach, the three-way array is unfolded into a single two-way array. There are several unfolding choices, which differ in the number of process variables lagged in time (the so-called LMVs or lags): batch-wise (BW) unfolding,[118, 119] variable-wise (VW) unfolding[186] (also called observation wise unfolding (OWU)[187]), and batch dynamic (BD)[35, 100, 172] – see Figure 6.1a–c, respectively. According to the research work published by Kourti,[93] only BW (Eq. 6.1) and VW (Eq. 6.2) unfoldings are of interest for the treatment of batch process data considering the variability to be modeled:

$$\underline{\mathbf{X}}(I \times J \times K) \Rightarrow \mathbf{X}(I \times JK) \quad (6.1)$$

$$\underline{\mathbf{X}}(I \times J \times K) \Rightarrow \mathbf{X}(IK \times J) \quad (6.2)$$

Nonetheless, the BD unfolding allows practitioners to capture the process dynamics up to a certain order when prior process knowledge is available for modeling specific time-varying dynamics for more accurate modeling. This unfolding can be seen as a generalization of the traditional unfolding procedures[28]: if no lags are added, the resulting two-way array is the same as the one after VW unfolding – i.e., $\mathbf{X}\ (IK \times J)$; if all possible lags are added, the resulting two-way array is the same as the one after BW unfolding – i.e., $\mathbf{X}\ (I \times JK)$. The addition of a certain number of lags depends on the order of the dynamics that need to be modeled, which is intrinsically related to how the correlation structure changes throughout the batch run, i.e., the way process variables are related to each other and over time.[28]

The batch dynamic unfolding can be expressed using the following notation:

$$\mathbf{X} = \text{unfold}(\underline{\mathbf{X}}, n) \equiv \underline{\mathbf{X}}^{(n)} \quad (6.3)$$

where n stands for the number of lags included as variables and:

$$n = \{k - 1 \quad : \quad k \in \{1, 2, \ldots, K\}\} \quad (6.4)$$

Using this notation, the BW unfolding is herein expressed as:

$$\mathbf{X} = \underline{\mathbf{X}}^{(K-1)} \quad (6.5)$$

and the VW unfolding as follows:

$$\mathbf{X} = \underline{\mathbf{X}}^{(0)} \quad (6.6)$$

6.2.1 Batch-wise Unfolding

The BW unfolding transforms the data of a complete batch, including all process variables and sampling times, into a single object (row or observation) of the BW unfolded two-way array $\underline{\mathbf{X}}^{(K-1)}$, which is illustrated in Figure 6.1a. The dimension of this array is $I \times JK$. Thus, the covariance matrix $\mathbf{C}_X$ computed from the batch-unfolded data is of dimension $JK \times JK$. The structure of this matrix is shown in Figure 6.1a.[1]

1 Only the auto-covariances and lagged cross-covariances of orders 1 and 2 are shown for simplicity.

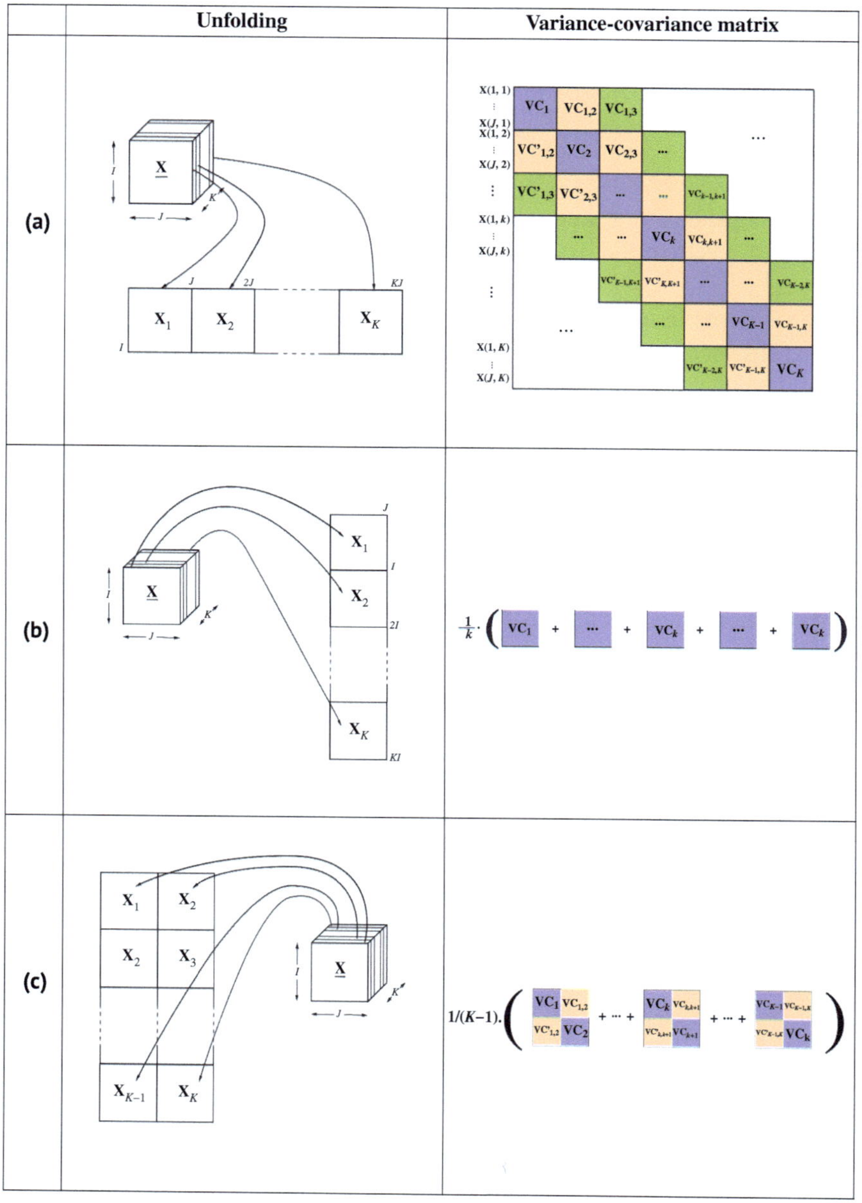

Figure 6.1 Single-model approaches to transform the three-way array into a two-way array with their associated variance–covariance matrices: (a) BW, (b) VW, and (c) BD with one LMV.

The instantaneous relationships of the process variables are captured by the sub-matrices $\{\mathbf{VC}_1, \mathbf{VC}_2, \ldots, \mathbf{VC}_K\}$ located on the main diagonal of the variance-covariance matrix, which represent the variances and instantaneous cross-covariances of the variables at every sampling time. The dynamics in the data are represented by the relationships in sub-matrices $\mathbf{VC}_{k,k+d}$ for $k = 1, \ldots, K - d$, which represent the auto-covariances and lagged cross-covariances of order d, i.e., the dynamic relationships of the process variables. Hence, we can conclude that BW unfolded models incorporate the linear dynamics of the process. Furthermore, these models are able to capture time-varying relationships among variables, since submatrices VC_k and $V_{k,k+d}$ for two different values of k – e.g., 1 and 10 – are placed in different parts of the covariance matrix, and hence, can be modeled independently. Since all possible relationships among variables and in time are included in $\mathbf{C}_X$, the BW unfolding allows us to study all the possible time-varying dynamics through the calibration of a PCA (or PLS) model on this covariance matrix, despite some authors' assertions.[35, 145]

A drawback of these models is that the flexibility to capture changing dynamics is at the expense of a heavy over-parametrization. $\mathbf{VC}_X$ is $JK \times JK$ and so each PC has JK loadings which are individually identified with I observations, when I is typically low. Consequently, a low stability – also known as high parameter uncertainty – is expected in these models. Uncertainty is related to parameter stability, and mainly depends on two factors, as already commented in Chapter 2. First, precise identification relies on an informative dataset and on the availability of a sufficiently large calibration dataset. Second, the more different the sum of squares (SS) captured by each PC, the more stability in the model parameters.[85] The next example illustrates how parameter stability is affected when calibrating a PCA model on BW unfolded data.

Hereafter, to proceed with the comparison of all the unfolding methods in terms of parameters stability, including preprocessing and model parameters, the normalized squared difference (NSD) is used. NSD is computed as follows:

$$\mathrm{NSD}_{\theta} = \sum_{j=1}^{J}\left(\frac{\theta_j^{(1)}}{|\boldsymbol{\theta}^{(1)}|} - \frac{\theta_j^{(2)}}{|\boldsymbol{\theta}^{(2)}|}\right)^2 \tag{6.7}$$

where $\theta_j^{(1)}$ and $\theta_j^{(2)}$ correspond to the jth value in the parameter vectors $\boldsymbol{\theta}^{(1)}$ and $\boldsymbol{\theta}^{(2)}$ for the first and second dataset, respectively. To make the NSD values of the loadings comparable across approaches, two factors need to be taken into account in the estimation: the number of PCA models and the number of parameters. The size of the loading vectors in each model depends on the number of lags added as new variables. To make all the models comparable, the NSD values are estimated as an average of the NSD values calculated on the loadings corresponding to each sampling time k (i.e., $\mathrm{NSD}_{\theta} = \sum_{k=1}^{K} \mathrm{NSD}_{\theta_k}/K$, where NSD_{θ_k} is assessed by the previous equation. When including lags, an exception is made for BW models, data from a specific sampling time point are used more than once to fit parameters in the same or different submodels. When this occurs, parameters in the form of lags are not considered computing the NSD values. To properly estimate the NSD values in loadings, the sign change of loadings due to the rotational ambiguity in PCA is taken into account. For this purpose, each loading vector $\mathbf{p}_a$ is corrected by the sign of the absolute maximum loading.

Example 6.1: Batch-wise unfolding

This subsection presents an example to study the parameter stability of preprocessing and PCA model parameters associated with the BW unfolding of batch data.

As starting point, let us load the NOC batch data synchronized with the relaxed greedy time warping (RGTW) algorithm for two datasets under study (see Chapter 4).

First, we preprocess the three-way data arrays using the trajectory centering and scaling (trajectory C&S) method. Then, we pass the three-way arrays *cn_alg* or *cn_alg2* as the first input, along with value 2 to indicate trajectory C&S[2] method. For the first dataset, we execute the following instruction:

```
[cn_ap,m,sd] = preprocess3D(cn_alg,2);
```

The next step is to apply BW unfolding as illustrated in Figure 6.1a. This process involves transforming the data of a complete batch, including all the process variables measured at different sampling times, into a single row or observation, in an unfolded two-way array. The ||**unfold**|| routine performs this unfolding in such a way that the measurements corresponding to the first sampling time are located in the first columns of the unfolded array, and subsequent measurements are placed to the right. To proceed with the BW unfolding, the second parameter of the ||**unfold**|| function has to be set to the length of the batch (minus 1) or any higher number, including the value *Inf*, which allows us to add all the possible lags.

```
cn_apu = unfold(cn_ap,Inf);
```

Note that it is sometimes preferable to arrange all the measurements corresponding to the same process variable together, for enhanced visualization of the model parameters and statistics. The following code rearranges the data matrix as described.

```
for i=1:J
    cn_apub = [cn_apub cn_apu(:,i:J:end)];
end
```

Once batch data have been preprocessed and BW unfolded, we calibrate a PCA model using 10 PCs. Recall that the MATLAB® native function "**pca**" from the Statistics and Machine Learning Toolbox applies mean centering. Since we have previously applied trajectory C&S, we can explicitly indicate not to preprocess the data, by setting the input parameter "Centered" to false or 0.

```
[p,t] = pca(cn_apub,'NumComponents',10,'Centered',0);
```

This process is afterward repeated for the second dataset *cn_alg2*, following the instructions available in the MATLAB live script called "ex6_1.mlx." After performing the analysis for the two datasets, we can address the question of what is the amount of calibration data which is enough to accurately identify the parameters. In Figure 6.2, the preprocessing information (i.e., means and standard deviations)

---

2 For detailed information on trajectory C&S and its computation, the reader is referred to Chapter 5 and the MATLAB code of the function ||**preprocess3D**||.

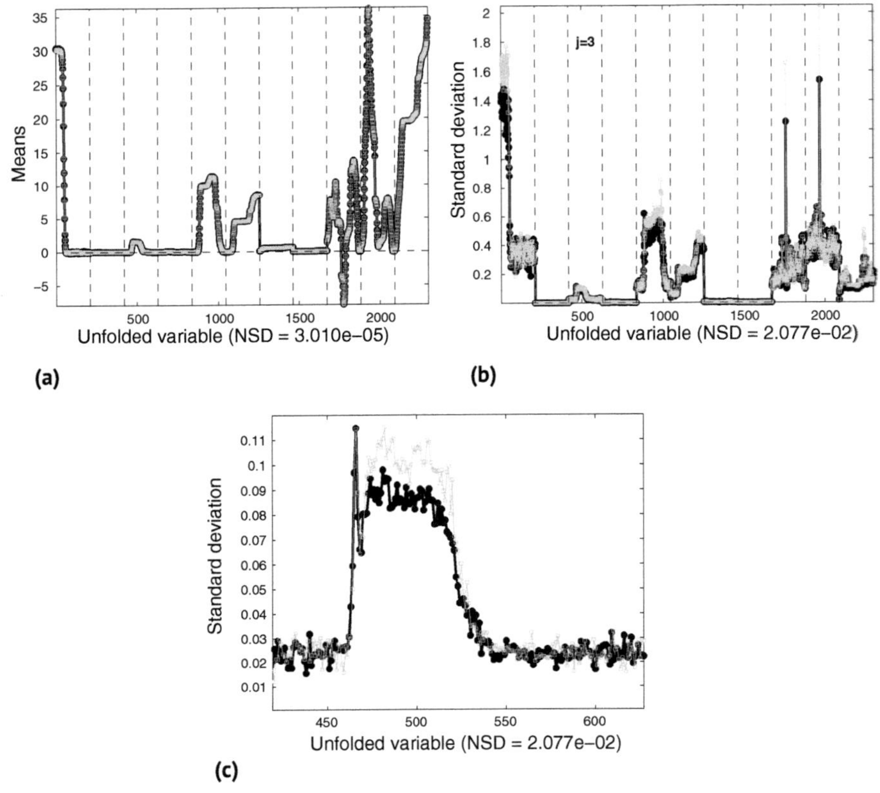

Figure 6.2 BW unfolding and trajectory C&S. Comparison of the preprocessing parameters ([a] means and [b] standard deviations) obtained from the two simulated datasets BW unfolded, after applying trajectory C&S, including (c) a zoom-in view of SD for variable #3 (acetaldehyde concentration). NSD: normalized squared differences between the average and standard deviations vectors of the two simulated datasets. *Source*: Reproduced from Ref.[64], with permission of John Wiley & Sons.

corresponding to the two datasets generated is compared. At first glance, the preprocessing parameters identified seem to be identical. Nonetheless, the zoom performed in Figure 6.2c shows that there are slight differences. The reason for this is that a high number of means ($J \cdot K$) and standard deviations ($J \cdot K$) is identified by exclusively using I batches, which is typically a low number compared to the number of estimated parameters. For instance, in the present example, $J \cdot K = 2090$ means and standard deviations are computed from $I = 30$ batches. The parameter uncertainty can also be checked using the NSD values computed for the means and the standard deviations: 3.01×10^{-5} and 2.077×10^{-2}, respectively. As noticed, there is variability in the preprocessing statistics between the two datasets, being lower in the mean than in the standard deviations. The fact that the uncertainty in the mean is lower than that in the standard deviation is indeed a systematic result. Hence, the resulting uncertainty in the standard deviation is due to the accumulation of variability in both types of preprocessing parameters.

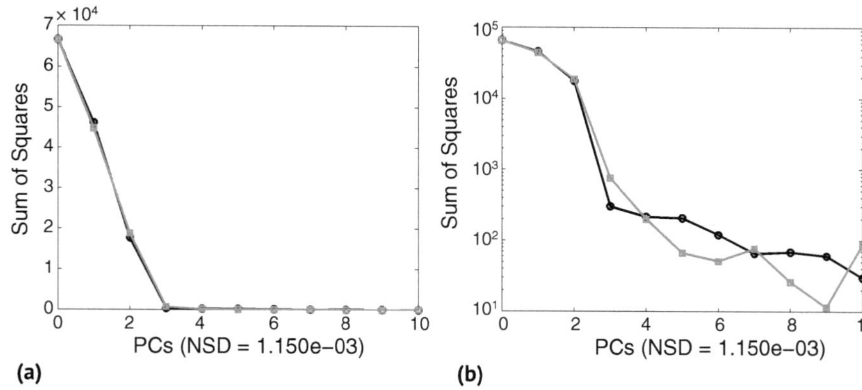

Figure 6.3 BW unfolding and trajectory C&S. Explained SS vs #PCs extracted in BW unfolding from the two simulated datasets. Note that PC #0 corresponds to the SS remained after applying trajectory C&S on batch data. (a) Linear scale and (b) logarithmic scale. NSD: normalized squared differences between the SS vector captured by each PC of the two simulated datasets. *Source*: Reproduced from Ref.[64], with permission of John Wiley & Sons.

We have also discussed previously that if the SS extracted in each PC is different enough in comparison to subsequently extracted PCs, low uncertainty in the parameters estimation is expected for a large calibration dataset. In some situations and for some applications, it is not a problem to have several PCs with a similar amount of SS captured. All of them can be included in the PCA model, or otherwise discarded and left in the residuals. Nonetheless, it is important to be aware of the uncertainty introduced in the loadings when this occurs. For this reason, it is always recommended to have a look at the SS captured by each PC. Figure 6.3 shows the plot of the explained SS versus the number of PCs extracted for the current example, assuming the application of BW unfolding and trajectory C&S. For the sake of visualization, both the linear and logarithmic scales are presented. As observable in the figure, the SS captured by PC#1 ($SS_1 \approx 4.500 \times 10^4$) explains a high percentage of the total sum of preprocessed data, i.e., SS_0, approximately 70%, and differs enough from the SS captured by PC#2 ($SS_2 \approx 1.800 \times 10^4$). Consequently, the loadings of the first PC are expected to be reasonably stable. Note that the SS captured from PC#3 onward are similar, and therefore, their corresponding loadings are not expected to be stable. In the current study, we only focus on those PCs which are expected to be stable in order to draw conclusions about the effect of applying a specific unfolding method on parameter stability. Hence, parametric uncertainty induced by a specific unfolded data structure is distinguished from the parameter stability due to PCs with similar captured variance, which is expected to affect the PCA models independently of the unfolding method of choice.

Figure 6.4 illustrates the two loading vectors corresponding to the first PC obtained for the two datasets generated. Note that the uncertainty in the preprocessing estimation is propagated in the identification of the PCs. In particular, the NSD value corresponding to the first loading vector of both datasets is equal to 1.524×10^{-1}, denoting an increasing uncertainty with respect to the preprocessing parameters. Furthermore, each PC contains $J \cdot K$ parameters, the same number of means or

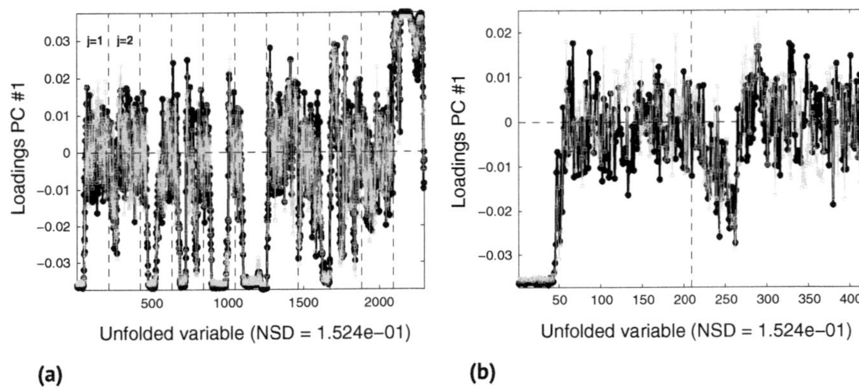

Figure 6.4 BW unfolding and trajectory C&S. Comparison of the loading vector corresponding to the first PC obtained from the two simulated datasets BW unfolded, after applying trajectory C&S: (a) complete loading vectors and (b) zoom corresponding to variables #1 and #2. NSD: normalized squared differences between the first loading vector of the two simulated datasets. *Source*: Reproduced from Ref.[64], with permission of John Wiley & Sons.

standard deviations previously estimated. The parameters are, again, estimated from I observations each. It is clear that there is a parallelism between trajectory C&S, and the components of a LVM after BW unfolding from the point of view of uncertainty estimation. Upon a closer look at Figure 6.4b, the loadings corresponding to the glucose concentration (variable #1) and pyruvate concentration (variable #2) are shown. Several loadings present different signs in the two datasets. Nonetheless, most of this variability is due to noise, since most loadings take values around zero (e.g., see Figure 6.4b from the 60th to the 209th loading and from the 260th to the 418th loading belonging to variables #1 and #2, respectively). Despite the fact that with BW unfolding a very complete modeling structure can be estimated,[31] the noisy loadings shown in Figure 6.4 suggest model overparametrization (i.e., overfitting). In any case, important parts are captured. Take for instance, the loadings of high magnitude in the interval [1, 50] are reflecting the high autocorrelation of the first variable (i.e., glucose concentration) during that period in the aligned datasets (see Figure 6.5).

6.2.2 Variable-wise Unfolding

The VW unfolding, which generates the array $\underline{\mathbf{X}}^{(0)}$, is illustrated in Figure 6.1b. This approach transforms the data collected at each sampling time of a batch as an object of the unfolded two-way array. This array is $IK \times J$, typically for IK some orders of magnitude higher than J, which is a much favorable situation for parameter identification than in BW models. The $J \times J$ covariance matrix $\mathbf{C}_X$ resulting after VW unfolding is shown in Figure 6.1b. This matrix is composed of the sum of the sub-matrices $\{\mathbf{VC}_1, \mathbf{VC}_2, ..., \mathbf{VC}_K\}$ weighted by the factor $1/K$. Note that these matrices are those in the diagonal of the variance-covariance matrix after BW unfolding, as shown in Figure 6.1a. Hence, only the variances and instantaneous cross-covariances of the process variables are present in the global covariance and

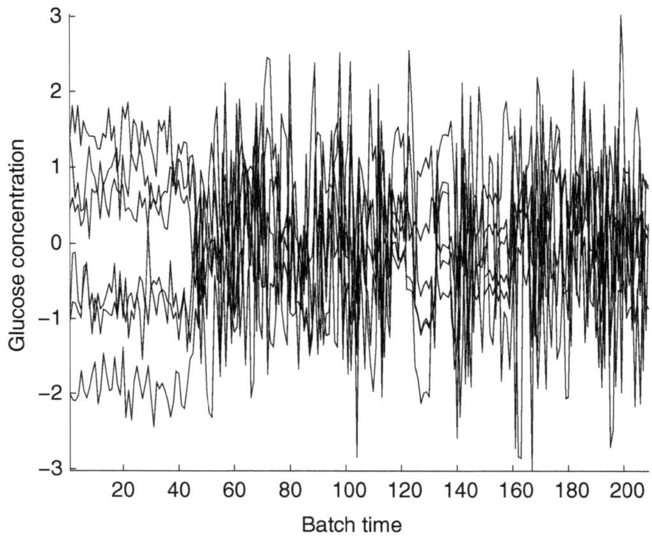

Figure 6.5 BW unfolding and trajectory C&S. Trajectory of glucose concentration (variable #1) after trajectory C&S in some of the batches of the second simulated dataset. *Source*: Reproduced from Ref.[64], with permission of John Wiley & Sons.

not the dynamics of the process. Furthermore, the global covariance is an average of the individual time covariances throughout the batch duration.[28] Consequently, this modeling approach assumes that 1) there are no dynamics (or at least that dynamics do not need to be modeled), and 2) the correlation structure of the process does not change over time. These are strong assumptions that are rarely valid in batch processes, where the process dynamics are often time-varying. Due to these strong assumptions not often met by actual data, the application of VW unfolding in this scenario may have a notable impact on prediction[29] and fault detection.[31]

Example 6.2: Variable-wise unfolding

In this example we study the influence of two preprocessing methods on parameter stability after calibrating a PCA model on VW unfolded data. The methods are variable centering and scaling (variable C&S), commonly associated with VW unfolding, and trajectory C&S, widely used with BW unfolding. Note that both preprocessing methods can be used indistinctly along with VW unfolding, but have different implications. Variable C&S focuses on studying deviation over the overall mean, while trajectory C&S focuses on the variability around the average batch trajectory.

Similarly to the experiment illustrated in the previous example, we start by loading the two datasets containing batch data synchronized by the RGTW algorithm. Afterward, we pass on the three-way array *cn_alg* as the first input, along with value 5 in this example to indicate the variable C&S method.

```
[cn_ap_vcs,m_vcs,sd_vcs] = preprocess3D(cn_alg,5);
```

The next step is to apply VW unfolding. This step transforms the data collected at each sampling time of a batch into an object or row of the unfolded two-way data array. Similarly to the example of the BW unfolding, we use the ||**unfold**|| routine, where the block of measurements belonging to the first sampling time is located in the first rows of the two-way unfolded array, and subsequent blocks of measurements corresponding to the rest of the sampling times are placed below each data block.

```
cn_apu_vcs = unfold(cn_ap_vcs,0);
```

Once variable C&S preprocessing and VW unfolding have been completed, we calibrate a PCA model using 10 principal components, indicating the MATLAB instruction below, suggesting that no further preprocessing is required, as explained in the previous example.

```
[p_vcs_vw,t_vcs_vw] = pca(cn_apu_vcs,'NumComponents',10,
    'Centered',0);
```

This procedure is afterward repeated for the second dataset under study.

For comparison purposes, the same steps are repeated using trajectory C&S. Instructions and the code can be found in the MATLAB live script called “ex6_2.mlx.”

Recall that a factor on which the parameter stability depends on the number of observations used in the estimation. In variable C&S, a total of J means and J standard deviations are identified using $I \cdot K$ observations. In this example, $J = 10$ and $I \cdot K = 6270$. Due to the fact that the parameters-to-observation ratio in trajectory C&S $\left(R_{\text{TCS}} = \frac{J \cdot K}{I} = \frac{2090}{30}\right)$ is much higher than in variable C&S $\left(R_{\text{VCS}} = \frac{J}{I \cdot K} = \frac{10}{6270}\right)$, the uncertainty in the estimation in the former is also higher than in the latter.

In Figure 6.6, the explained SS versus the number of PCs extracted for the current example assuming VW unfolding and variable C&S (see Figure 6.6a) and trajectory C&S (see Figure 6.6b) are shown. Note that the SS remaining after variable C&S for both datasets ($\text{SS}_0 = 6.896 \times 10^4$) is slightly higher than after trajectory C&S ($\text{SS}_0 = 6.667 \times 10^4$). This has nothing to do with stability, and it is due to the different types of preprocessing carried out. In the former, since the total variance is set to 1 for each process variable, the remaining SS is equal to $\text{SS}_0 = (I \cdot K - 1) \cdot J = (30 \cdot 209 - 1) \times 11 = 6.896 \times 10^4$ units. In the latter, since the variance is set to 1 for each variable and sampling time, the total variance is equal to $\text{SS}_0 = (I - 1) \cdot K \cdot J = (30 - 1) \times 209 \times 11 = 6.667 \times 10^4$.

Let us proceed as in the previous example, assessing the sum of squares captured by PC. First, in the case of VW with variable C&S shown in Figure 6.6a, the sum of squares captured by PC#1 ($\text{SS}_1 \approx 4.145 \times 10^4$) explains approximately 60% of the variance in data, and it is different enough compared to that captured by PC#2 ($\text{SS}_2 \approx 7.700 \times 10^3$). Consequently, the loadings of the first PC are expected to be stable. The sum of squares of PC#2, PC#3, and PC#4 seem to be quite similar, and therefore, their loadings may not be stable. Notice that this result is specific to the dataset at hand, and not a feature of the modeling and/or preprocessing method. Secondly, in the case of VW with trajectory C&S (Figure 6.6b), the sum of squares captured by PC#1 ($\text{SS}_1 \approx 5.100 \times 10^4$) explains approximately 75% of the variance in the data, and again, it is different enough compared to the sum of squares captured by PC#2 ($\text{SS}_2 \approx 7.500 \times 10^3$). As a consequence, the loadings of the first PC are expected to be stable. Note that the sum of squares captured from PC#2 onward are also similar in this case,

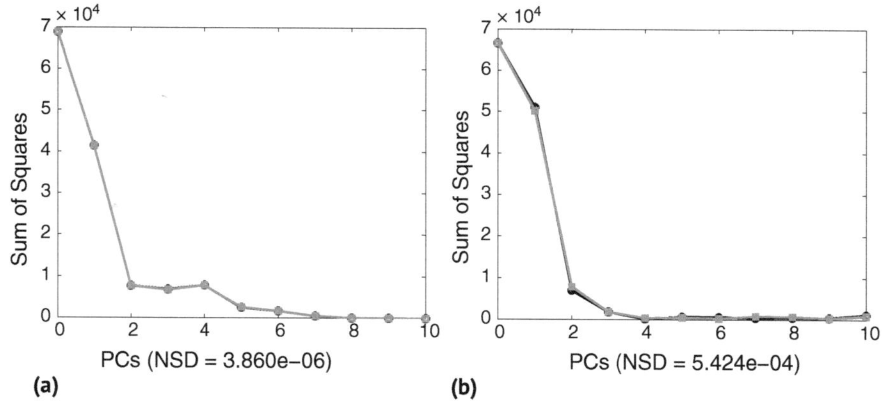

Figure 6.6 VW unfolding. Explained SS vs #PCs extracted in VW unfolding from the two simulated datasets. Note that PC #0 corresponds to the SS remained after applying (a) variable C&S and (b) trajectory C&S. NSD: normalized squared differences between the SS vector captured by each PC of the two simulated datasets. *Source*: Reproduced from Ref.[64], with permission of John Wiley & Sons.

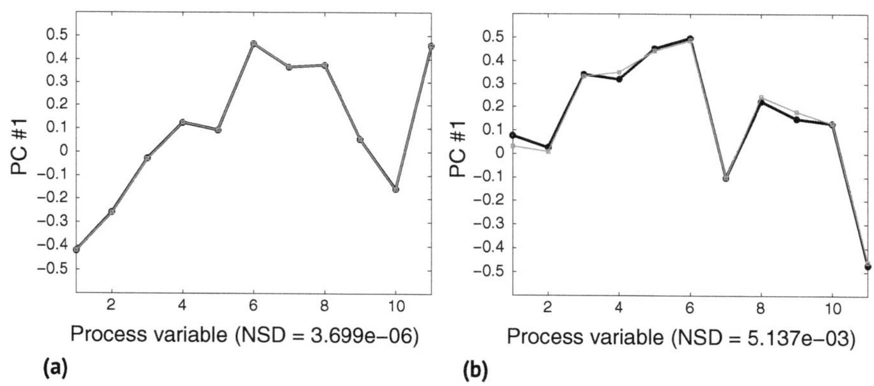

Figure 6.7 VW unfolding. Comparison of the loading vector corresponding to the first PC obtained from the two simulated datasets between (a) variable C&S and (b) trajectory C&S. NSD: normalized squared differences between the first loading vector of the two simulated datasets. *Source*: Reproduced from Ref.[64], with permission of John Wiley & Sons.

and so their corresponding loadings are not expected to be stable. Furthermore, the uncertainty measured in the residual sum-of-squares by each PC of each VW model via the NSD values (NSD = 3.860×10^{-6} and NSD = 5.424×10^{-4} for VW unfolding with variable C&S and for VW unfolding with trajectory C&S, respectively) confirms that variable C&S generates more stable estimates than trajectory C&S.

In order to compare the stability of the first PC, the corresponding loadings for both preprocessing methods are shown in Figure 6.7. In terms of NSD, the uncertainty observed in the loadings after variable C&S (NSD = 3.699×10^{-6}) is approximately three orders of magnitude lower than after trajectory C&S (NSD = 5.137×10^{-3}). This uncertainty is propagated from a similar difference in the uncertainty of the preprocessing parameters. Hence, stability of the loadings of PC#1 in variable C&S is higher than in trajectory C&S.

The results in terms of stability should be interpreted with care and in connection with other features of the models, as those discussed in the papers.[28, 29] It should be noted that low parameter stability does not guarantee the model is adequate. Note that the variability of interest in batch MPSC (BMSPC) is the deviation of a batch from the common trend reflected on the average trajectory of the process. When the average trajectory is not extracted in the preprocessing, as in variable C&S, the associated variability remains in the data. If the data are subsequently VW-unfolded, that specific variability turns into nonlinear relationships which cannot typically be captured with a linear model, such as PCA. Therefore, VW after variable C&S is not suited to capture the variability of interest in BMSPC.

6.2.3 Batch Dynamic Unfolding

Chen and Liu[35] proposed a different way of arranging the three-way data array in two dimensions for online monitoring. We call this unfolding, followed by PCA or PLS modeling, BD. The equivalence of this method to the VW unfolding with additions of lags was demonstrated.[27] In this unfolding approach, an augmented two-way data array $\mathbf{X}_{i,n}((K-n)\times J\cdot(n+1))$ is generated for every batch, including the n immediate lags:

$$\mathbf{X}_{i,n} = \begin{bmatrix} \mathbf{x}_{i,1}^T \cdots \mathbf{x}_{i,n}^T\ \mathbf{x}_{i,n+1}^T \\ \mathbf{x}_{i,2}^T \cdots \mathbf{x}_{i,n+1}^T\ \mathbf{x}_{i,n+2}^T \\ \cdots \\ \mathbf{x}_{i,K-n}^T \cdots \mathbf{x}_{i,K-1}^T\ \mathbf{x}_{i,K}^T \end{bmatrix} \tag{6.8}$$

where $\mathbf{x}_{i,k}^T(1\times J)$ is the measurement vector of the J variables for batch i and sampling time k.

Then, a *pseudo-covariance* matrix $\mathbf{pS}_{i,n}$ is calculated for every $\mathbf{X}_{i,n}$ (Eq. 6.9). The term "pseudo" is used here because Chen and Liu subtract the average trajectory from the data, so that the mean vector of $\mathbf{X}_{i,n}$ does not have to be exactly zero.[35]

$$\mathbf{pS}_{i,n} = \frac{(\mathbf{X}_{i,n})^T(\mathbf{X}_{i,n})}{K-n-1} \tag{6.9}$$

Finally, the *pseudo-covariance* matrices are averaged obtaining matrix $\mathbf{S}_{avg,n}$ (Eq. 6.10), which is the same matrix but a constant factor (the denominator) as the covariance matrix of the VW unfolded two-way data array when including n immediate lags as variables.

$$\mathbf{S}_{avg,n} = \frac{(K-n-1)\sum_{i=1}^{I}\mathbf{pS}_{i,n}}{J(K-n)} \tag{6.10}$$

Figure 6.1c shows an example of the batch dynamic unfolding method with the addition of one LMV along with the associated covariance matrix. From this figure it is easy to understand the effect of the addition of lags. Dynamics of order d are built in the covariance (and corresponding LVM) by matrices $V_{k,k+d}$. The more lags added, the more dynamic information included in the model. This unfolding method can be seen as a generalization of the traditional unfolding procedures: if no LMV is added, the resulting two-way data array is the same as the one after VW unfolding; if all possible lags are added, the resulting two-way data array is the same as the one after BW unfolding. As stated by Camacho and Picó,[26] depending on the nature of the process, part of the dynamic information may be negligible, and it could be advantageous not to incorporate it into the model. Thus, identifying the optimal number of lags from the data of a process is, *a priori,* a more powerful modeling approach than using an extreme case, like in BW or VW unfolding. Nevertheless, it should be noted that PCA models are neither predictive models nor autoregressive models. Thus, the introduction of dynamics within the PCA model does not affect its modeling performance to the same extent as for the other models, as it is shown in the following example.

The introduction of lags has a side effect apart from adding dynamic information. When the number of lags is low, auto-correlations and cross-correlations are obtained independently of the precise sampling time. Nonetheless, as it happens for VW unfolded models, modeling with a reduced number of lags assumes lags included in the model, the less the number of objects – rows – in the unfolded array – leading to a two-way array $\mathbf{X}$ of $I \cdot (K - n)$ rows. Therefore, by adding lags, the interval where the correlation structure is imposed to be constant is reduced, and the model is more capable of capturing changes in the dynamics. At the same time, the dynamics built in the model are more dependent on the precise sampling time. Generally speaking, a batch dynamic model with n lags assumes the dynamics of the process can be considered time-invariant for a $K - n$ sampling times.

Although batch dynamic unfolding provides the possibility to capture time-varying correlation structures by adding a sufficient number of lags, this approach is not optimal in all situations since it may lead to over-parameterized models. Take the extreme example of a process in which there is an absence of time-varying dynamics and only static relationships among variables exist. This process can be modeled by BW unfolding the data and applying a bilinear LVM. This would be equivalent to adding all possible lags in batch-dynamic unfolding. Using that model, time-varying static relationships are effectively modeled by matrices VC_k in Figure 6.1a. Nonetheless, the rest of the covariance matrix is just noise, since there are no significant dynamic relationships in the data.

Example 6.3: Batch dynamic unfolding

The experiment with the two datasets is repeated here for batch-dynamic unfolded data with one LMV and trajectory C&S preprocessing. The MATLAB code to execute the workflow and generate the results of this example, which will be thoroughly discussed next, is available in the MATLAB live script called "ex6_3.mlx."

As a starting point, we preprocess the three-way data arrays using the trajectory C&S method. The next steps are identical as in Example 6.1, except for the input parameters of routine ||**`unfold`**||, so that the second input is set to 1 as shown below:

```
cn_apu_bd = unfold(cn_ap,1);
```

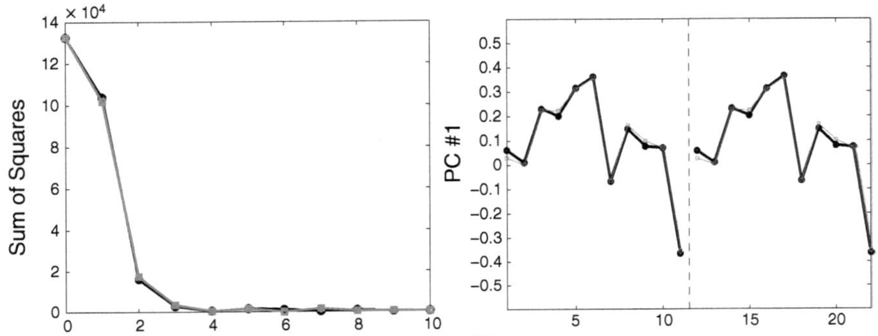

(a) **(b)**

Figure 6.8 BD unfolding and trajectory C&S: (a) explained SS vs #PCs extracted and (b) first loading vector in BD unfolding with one LMV from the two simulated datasets. Note that PC #0 corresponds to the SS remained after applying trajectory C&S on batch data. NSD: normalized squared differences between (a) the sum of squares vector captured by each PC and (b) the first loading vector of the two simulated datasets. *Source*: Reproduced from Ref.[64], with permission of John Wiley & Sons.

Figure 6.8 shows the explained sum of squares versus the number of PCs and the loading vectors corresponding to the first PC for the two datasets. These results are quite similar to those obtained for VW unfolding and trajectory C&S. Hence, Figures 6.8a and 6.6b present a very similar shape, with the main difference that the former doubles the latter in explained sum-of-squares. This is the logical consequence of doubling the number of variables by adding one lag, so that each observation appears twice in the unfolded data array. Also, Figures 6.8b and 6.7b essentially present the same relationships among variables, but again, the former shows these relationships twice. Concerning the loading stability, this approach yields an intermediate uncertainty level, as shown with the NSD, between VW and BW unfolding. In particular, variability in BD is lower (NSD $= 1.019 \times 10^{-2}$) than in BW after trajectory C&S (NSD $= 1.524 \times 10^{-1}$), and higher than in VW after trajectory C&S (NSD $= 5.137 \times 10^{-3}$) and after variable C&S (NSD $= 3.699 \times 10^{-6}$). This result is expected since BD is a generalization of VW and BW (its parameters-to-observations ratio is higher than VW, but lower than BW). Figure 6.8b also shows that the autocorrelation in the data is so high that the loadings for one variable and its lagged version are almost identical.

6.3 *K*-models Approach

In the single-model approach, one bilinear model is generated for the whole process. In contrast, the K-models approach is based on generating as many bilinear models as there are sampling times in a batch. Several proposals can be found in the literature, which differ in the data used in the generation of the submodels. If each submodel exclusively incorporates measurements collected at the current sampling time, then we call it a local model[129] – i.e., $\mathbf{X}$ $(I \times J)$ (see Figure 6.9). Data are split according to the following equation:

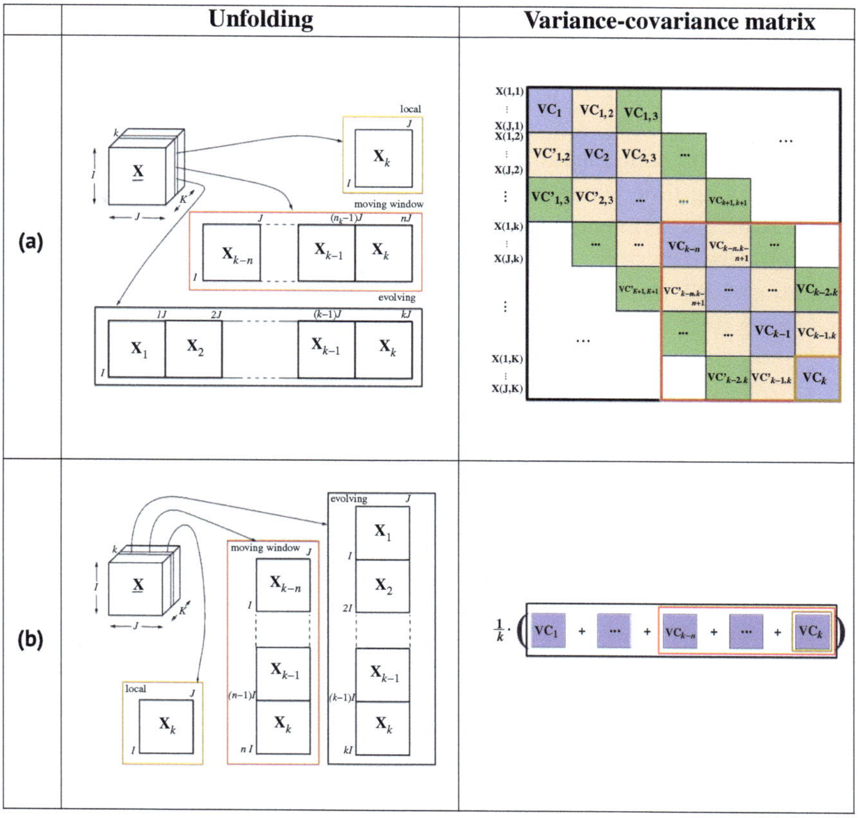

Figure 6.9 *K*-models approaches with all the possible lags added as (a) new variables and as (b) new observations with their associated variance–covariance matrices: local (yellow), moving window (red), and evolving (black).

$$\mathbf{X} = \{\mathbf{X}_k : k = 1, ..., K\} \tag{6.11}$$

To incorporate dynamics in the models, a straightforward alternative to local modeling is evolving modeling,[129] where all the possible lags at the kth sampling time point are included as additional variables[28]:

$$\mathbf{X} = \{\underline{\mathbf{X}}_{1:k}^{(k-1)} : k = 1, ..., K \tag{6.12}$$

The notation we use in Eq. (6.12) reflects, in the subindex, the sampling times incorporated in the two-way data array used to fit each submodel – i.e., in this case, from the first sampling time to the kth time, $\mathbf{X}$ ($I \times kJ$). In the superindex we specify the unfolding approach, like in the single model approaches. We can see that the evolving modeling approach only considers BW unfolding. Yet, this approach could be combined with any other unfolding method.

Special cases of evolving models are uniformly weighted moving window (UWMW)[102] and exponentially weighted evolving window (EWEW) models, which are used when not all past information (lagged information) is of interest or has the same importance in bilinear modeling.[28] UWMW models are based on modeling the information contained in a window of width n, i.e., the measurements collected at the kth current sampling time with those of the immediate n lags:

$$\mathbf{X} = \{\underline{\mathbf{X}}_{k-n:k} : k = 1, \ldots, K\} \tag{6.13}$$

The information contained in the window can also be seen as a local model at the kth sampling time where n lags are included as variables – i.e., $\mathbf{X}\,(I \times nJ)$ – (see data structure for moving window models in Figure 6.9a) or observations – i.e., $\mathbf{X}\,(nI \times J)$ – (see data structure for moving window models in Figure 6.9b).

In contrast, EWEW submodels include all the lagged measurements to the current kth sampling time, but not all are given the same importance:

$$\mathbf{X} = \{\underline{\mathbf{X}}_{1:k} \odot \underline{\mathbf{W}}_{1:k} : k = 1, \ldots, K\} \tag{6.14}$$

where $\underline{\mathbf{W}}_{1:k}$ is a weighting matrix and $\odot$ stands for the Hadamard (element to element) product. $\underline{\mathbf{W}}_{1:k}$ is often constructed according to an exponentially decreasing value, the *forgetting factor* $\lambda_k \epsilon [0, 1]$, so that each measurement loses importance as the process progresses. The weight of the measurement vector collected at time $k - d$, for the generation of the submodel at time k, is $(\lambda_k)^d$, where the weight of the current measurements is always $(\lambda_k)^0 = 1$. This is equivalent to a local model at the kth sampling time when all the possible lags are added either as variables – i.e., $\mathbf{X}\,(I \times kJ)$ – (see data structure for evolving models in Figure 6.9a) or observations – i.e., $\mathbf{X}\,(kI \times J)$ – (see data structure for evolving models in Figure 6.9b), and exponentially weighted. With this forgetting factor, the measurements are losing importance over the batch duration and their contribution to the covariance matrix is downweighted.[28] This can be problematic for processes where the final quality of the batch is strongly dependent on phenomena at the beginning of the batch (see Refs.[93] and[118] for a discussion on this topic in emulsion polymerization).

One of the advantages of these K-model approaches is that they are capable of capturing varying dynamics of order n when n lags are added as new variables (see Figure 6.9a), in contrast to when they are added as new observations (see Figure 6.9b). Since there is one model for every sampling time, changing – dynamic or instantaneous – relationships are modeled. In evolving models, all possible dynamics until the current time point k are included while the process evolves, whereas no dynamics are captured by local models, only the instantaneous correlations at the kth sampling time (see Figure 6.9). Moving Window models are an intermediate case. In UWMW models where lags are added as observations, variances and instantaneous cross-covariances of the process variables within the selected window of samples of width n are captured. When lags are added as variables, the process dynamics of the n sampling times are captured. In EWEW models, the same dynamics are captured as in UWMW models for the two versions, with the difference that these relationships are weighted by the factor $(\lambda_k)^d$. The main drawback of the K-models approach is the generation and maintenance of a high number of submodels.

Example 6.4: *K*-models

In this example, the two datasets are used to illustrate how the parameters stability varies with the different types of K-models previously discussed.

Prior to the calibration of any K-model, we preprocess the three-way data array using the trajectory C&S method.

Local models are calibrated for each kth sampling time, resulting in K PCA models. To illustrate the parameter stability of the local models, as well as other K-model approaches, we only extract 1 PC for each submodel.

```
for k=1:K
    p_local{k} = pca(squeeze(cn_ap(k,:,:))',
    'NumComponents',1,'Centered',0);
end
```

In another example, a uniform weighting function is used – so-called UWMW. For the sake of simplicity and comparison with other K-models, we calibrate single PCA models (local models) for each possible sampling time incorporating only dynamics of order 1 – i.e., one LMV. This means that K-1 UWMW models of order 1 are calibrated. For this purpose, note that we make use of the ||**unfold**|| routine to add one LMV, as previously done with the example of the batch dynamic unfolding.

```
for k=2:K
    p_uwmw_var{k-1} = pca(unfold(cn_ap(k-1:k,:,:),2),
    'NumComponents',1,'Centered',0);
end
```

In a second example of UWMW using a uniform weighting function, the LMV is added as new observations. Therefore, the instantaneous relationship, and not the dynamics, are studied. Similarly to the previous example, we calibrate $K-1$ PCA models, and place data for J process variables at K sampling time below the data corresponding to the $K-1$ sampling time, resembling a VW unfolding for two consecutive sampling times. This means that $K-1$ UWMW models with LMV are added as new observations are calibrated.

```
for k=2:K
    p_uwmw_obs{k-1} = pca(unfold(cn_ap(k-1:k,:,:),0),
    'NumComponents',1,'Centered',0);
end
```

EWEW models do not give the same importance to all the lags, but each measurement is given an exponentially decreasing value, also called the forgetting factor λ, to lose importance as the process progresses. As explained earlier, using a forgetting factor, the measurements lose importance over the batch duration and their contribution to the covariance matrix is down-weighted. The following code calculates an exponential function defined by the number of K sampling times of the two simulated datasets.

```
expinc = 0.03;
expv=exp(0.01:expinc:(K-1)*expinc+0.01)./exp((K-1)*expinc
   +0.01)';
```

For illustration purposes, we calibrate EWEW models using the exponential function with λ equal to 0.97. For this purpose, we take $1-\lambda$ as a weighting factor, which is the first exponential function calculated and stored in the array called 'expv' above. Additionally, we add one LMV as new variables.

```
p_ewew_vars{k} = pca(unfold(w.*cn_ap(1:k,:,:),k-1),
'NumComponents',pc,'Centered',0);
```

In a second example of EWEW, we add the LMV as new observations, and use the same λ equal to 0.97. Similarly to UWMW, we calibrate as many PCA models as sampling times, but on this occasion, we place consecutive data for the two-way array corresponding to the kth sampling time below the two-way array corresponding to the $(k-1)$th sampling data, which resembles to a VW unfolding for two consecutive sampling times. This means that we calibrate K EWEW models with lags added as new observations.

```
for k=1:K
    w= ones(size(cn_ap(1:k,:,:)));
    for kk=1:k
      w(kk,:) = expv(K-k+kk,1);
    end
    p_ewew_obs{k} = pca(unfold(w.*cn_ap(1:k,:,:),0),
    'NumComponents',pc,'Centered',0);
end
```

The same procedure is repeated for the second dataset. For this purpose, the practitioner can find the complete sequence of instructions and code in the MATLAB live script called "ex6_4.mlx."

In Figure 6.10, the resulting loading vectors of the first PC for a) the local model, b) the UWMW model with one LMV in the variables, c) the UWMW model with one LMV in the observations, d) the EWEW model with lags in the variables and $\lambda = 0.97$, and e) the EWEW model with lags in the observations and $\lambda = 0.97$. All the models shown correspond to sampling time $k = 10$ in the datasets, and in all the cases data were trajectory C&S.

Essentially, the instantaneous relationships captured in the models are the same, i.e., the loading vector profiles are basically similar. Nevertheless, such consistency does not necessarily have to apply to other processes or numbers of LMV. In Figure 6.10, the NSD values between the loadings corresponding to both datasets are also included. In the approaches where 1 or all the possible lags are added as new variables, the NSD value is computed on the loading vector defined by the last J loadings (corresponding to the kth current sampling time) instead of all the loadings (as in the single-model approaches). This is done to compare the K-models approach with the rest of the approaches under study.

Comparing the addition of lags as new variables with the addition of lags as new observations both in UWMW and EWEW, the former presents higher uncertainty ($NSD_{UWMW} = 2.640 \times 10^{-1}$ and $NSD_{EWEW} = 1.715 \times 10^{-1}$) than the latter ($NSD_{UWMW} = 2.405 \times 10^{-1}$ and $NSD_{EWEW} = 6.858 \times 10^{-2}$) (see Figure 6.10b, d in comparison with Figure 6.10c, e). Hence, when lags are added as new variables, there is a negative effect in terms of parameter fitting, as a consequence of increasing the number of parameters to be estimated. This means that adding new parameters –

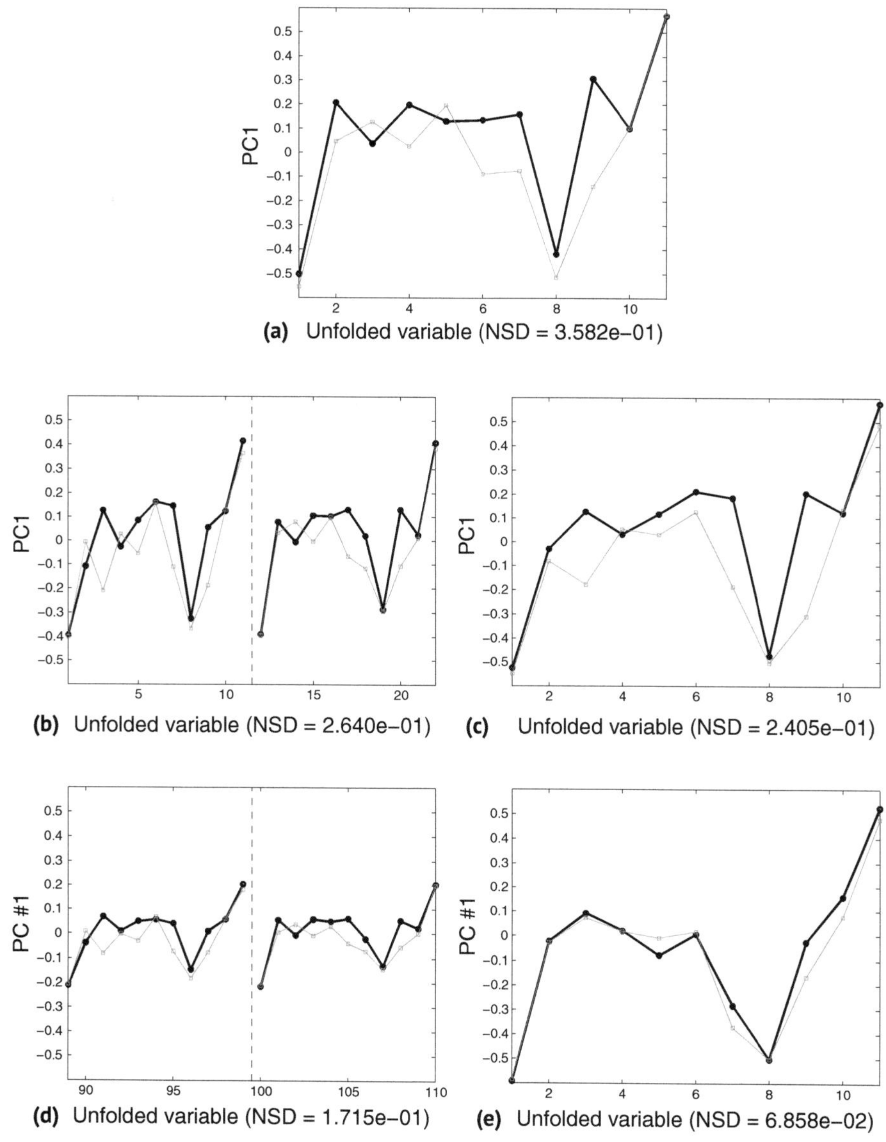

Figure 6.10 *K*-models and trajectory C&S. First loading vector for the two datasets at the kth sampling time: (a) local model, (b) UWMW model with one LMV in the variables, (c) UWMW model with one LMV in the observations, (d) EWEW model with lags in the variables and $\lambda = 0.97$ (only the loadings corresponding to the $k-$ and $(k-1)$th sampling time are shown for the sake of comparison) and (e) EWEW model with lags in the observations and $\lambda = 0.97$. In this approach, the NSD value is estimated as the average of the NSD values calculated at each k sampling time. *Source*: Reproduced from Ref.[64], with permission of John Wiley & Sons.

i.e., adding LMV as new variables – negatively affects the estimation of the parameters already in the model–those for instantaneous correlations (i.e., for the loadings corresponding to the current sampling time). On the other hand, adding LMV as new observations has a positive effect on the parameter stability, and it reduces the uncertainty on parameters estimation, as a consequence of increasing the number of observations to estimate each parameter.

For the present dataset and the metaparameters selected (number of lags and λ), it should be noted that the local models show a higher NSD value than the EWEW with lags in the variables and UWMW with one LMV in the variables approaches. This can be explained by the fact that autocorrelation and lagged cross-correlation also have a smoothing effect on loadings, which reduces the uncertainty. A similar effect can be observed by comparing the NSD value of the loadings corresponding to the first PC for BW (NSD $= 1.524 \times 10^{-1}$, see Figure 6.4) and local models. In both cases, a total of $J \cdot K$ parameters is estimated from the data of I batches. However, a BW PCA model takes into account the autocorrelation and lagged cross-correlation to improve the model estimation, while local PCA models do not. The result is a lower uncertainty in the former than in the latter. Therefore, the inclusion of lags as variables has a double and contradictory effect on the uncertainty. Generally speaking, the increase in the number of parameters augments the uncertainty. This happens unless that increase is justified by a high level of autocorrelation in the data. This supports the claim that the approach for transforming three-way into two-way arrays should be selected depending on the data at hand.[30]

6.3.1 Hierarchical-model Approach

The hierarchical-model approach is based on combining the past and current information at each sampling time with an adaptive hierarchical PCA model, i.e., adaptive hierarchical K-models (AHKM) (see Figure 1.2e).[131] First, a PCA model with one PC is fitted on the batch data corresponding to the first time slice $\widetilde{\mathbf{X}}_1$, leading to the scores and loadings $\mathbf{t}_{a1}$ and $\mathbf{p}_{a1}$ corresponding to the ath PC. The hierarchical part of the algorithm starts from the second sampling time. At sampling time k ($k = 2, \ldots, K$), the overall score vector $\mathbf{t}_{a(k-1)}$, which summarizes previous process variation up to the sampling time $k - 1$, is used as the starting vector to estimate the block scores $\mathbf{r}_{ak}$ derived from the two-way data array $\widetilde{\mathbf{X}}_k$ corresponding to a time slice k. Afterward, this score vector is weighted by the weighting factor λ and placed in the consensus matrix $\mathbf{R}_k$ together with the previous overall score vector $\mathbf{t}_{k-1}$.

The consensus matrix $\mathbf{R}_k$ is then used to calculate the overall score vector $\mathbf{t}_k$, which represents the total process variation up to the sampling time k. For this purpose, the weights vector $\mathbf{w}_k$ is calculated by multiplying the consensus matrix $\mathbf{R}_k$ by the overall score vector $\mathbf{t}_{a(k)}$. Then, the new overall score vector $\mathbf{t}_{a(k)}$ is calculated by multiplying the consensus matrix $\mathbf{R}_k$ by the weight vector $\mathbf{w}_k$. These estimations are repeated until the overall score vector $\mathbf{t}_k$ reaches convergence. This procedure is repeated for all the sampling times and all the PCs to extract. For further details about the algorithm, readers are referred to the original proposal published by Rännar et al.[131]

The weighting factor λ is similar in nature to the exponential weighting factor in an exponentially weighted moving average (EWMA) model. It is used to give relative importance to information collected at the current sampling time k with regard to past information. For high values of λ, the model is adapted quickly, while for low values of λ, the adaptation is slow. For λ values close to 0, the AHKMs approach uses memory of the past information, and therefore, this approach becomes similar to BW unfolding. As λ grows further than one, the AHKMs approach converges to the local K-models approach since the adaptive model does not use memory of any previous information.

The AHKM approach has been proved to work out well in terms of fault detection in a multistage context.[166] Nevertheless, one of the drawbacks of this approach is the need to estimate a model for each sampling time, making this difficult to implement for online applications. In addition, a proper weighting factor may not be found for the whole batch run due to transitions between phases. In this case, the use of hierarchical K-models would not be the best approach for process monitoring. The analysis of this approach from the covariance matrix is not trivial and is not carried out here.

Example 6.5: Adaptive hierarchical K-models

The two example datasets are used to show how varying different weighting factors influence parameter stability in AHKM models. For this study, we examine the extreme weighting factor λ values:

```
lambda_a = 0.2;
lambda_b = 50;
```

The next step is to apply BW unfolding.

```
cn_apu = unfold(cn_ap,Inf);
```

Afterward, we define an array of indices corresponding to all the lags added as part of the BW unfolding method to pass it on as an input parameters to the AHKM method.

```
indX = zeros(K,2);
for k=1:K
    indX(k,:)=[(k-1)*J+1 k*J];
end
```

Next, we execute the ||**aHPCA**|| routine, which calibrates an AHKM given: (i) a number of components (in this case PCs = 10) and (ii) λ values 0.2 and 50 as defined earlier.

```
[Pk1a1,Wk1a1,Tk1a1,Rk1a1]=aHPCA(cn_apu,indX,PCs,lambda_a);
[Pk2a1,Wk2a1,Tk2a1,Rk2a1]=aHPCA(cn_apu2,indX,PCs,
   lambda_b);
```

Note that the same procedure needs to be executed for the second dataset. To ease the calibration of all the AHKM and generate the results that are discussed in the next paragraphs, the corresponding code is available in the MATLAB live script called "ex6_5.mlx."

Firstly, let us study the identification of the PCA model parameters through the sum of squares captured for each PC. Figure 6.11 shows the explained sum of squares (SS) versus the number of PCs for an AHKM approach by using weighting factors $\lambda = 0.2$ (see Figure 6.11a) and $\lambda = 50$ (see Figure 6.11b). Recall that a λ close to 0,

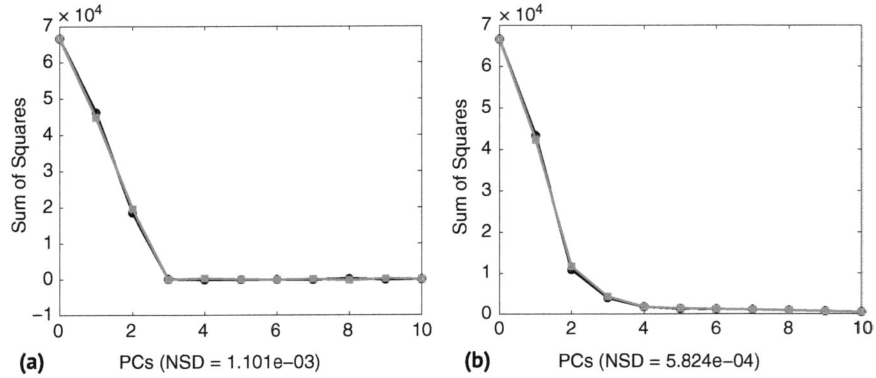

Figure 6.11 AHKMs and trajectory C&S. Explained SS vs #PCs extracted in the AHKMs with (a) $\lambda = 0.2$, and (b) $\lambda = 50$. Note that PC #0 corresponds to the sum of squares remained after applying trajectory C&S on batch data. NSD: normalized squared differences between the sum of squares vector captured by each PC of the two simulated datasets. *Source*: Reproduced from Ref.[64], with permission of John Wiley & Sons.

the AHKMs approach uses memory of the past information, and therefore, this approach becomes similar to BW unfolding. As λ grows beyond one, the AHKM approach converges to the local K-models approach, since the adaptive model down-weights the memory of any previous information. In the PCA with $\lambda = 0.2$ (see Figure 6.11a), the sum-of-squares captured by PC#1 ($SS_1 \approx 4.550 \times 10^4$) explains roughly 70% of the sum of squares and differs enough to the sum of squares captured by PC#2 ($SS_2 \approx 1.920 \times 10^4$). Hence, the corresponding loadings are expected to be stable, as in BW unfolding (see Figure 6.3 for comparison). As previously, from PC#3 onward, the explained sum of squares is similar, and consequently, their corresponding loadings are not expected to be stable. Regarding the AHKM model with $\lambda = 50$, a progressive decay of the explained sum of squares as a function of the number of tPCs can be observed, as shown in Figure 6.11b. The sums-of-squares captured by the first two PCs ($SS_1 \approx 4.290 \times 10^4$ and $SS_2 \approx 1.130 \times 10^4$) differ from each other enough to consider the loadings of the corresponding PCs stable. In contrast, the sum of squares captured from PC#3 onward are similar, so their corresponding loadings are not expected to be stable.

Aiming to further study the effect of the weighting factor in terms of parameter stability, AHKM was performed for the two datasets while adjusting the weighting factor from $\lambda = 0.1$ (roughly nonadaptive model) to $\lambda = 100$. The corresponding NSD values computed for the loadings of the first PC are shown in Figure 6.12. As noticeable, AHKM using $\lambda = 0.2$ reduces the differences found between the loadings of the first PC for the two datasets ($SD = 1.706 \times 10^{-1}$). Note that the value of λ that minimizes parameter stability may not be the same for different datasets and/or number of PCs. The differences between loadings obtained for the two datasets reach a platoon for $\lambda \geq 20$ (e.g., $NSD = 3.580 \times 10^{-1}$ for $\lambda = 50$), due to the fact that the adaptive hierarchical model approach converges to the classical local K-models approach – the curve of Figure 6.12 converges to the NSD value of Figure 6.10a. It apparently suggests that the lower the weighting factor, the more stable the model parameters

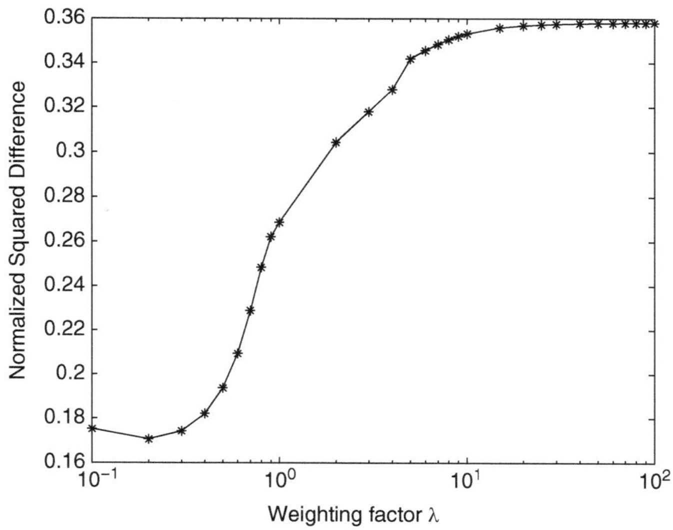

Figure 6.12 AHKMs and trajectory C&S. NSD between the first loading vector of the two datasets as a function of the weighting factor λ. *Source*: Reproduced from Ref.[64], with permission of John Wiley & Sons.

in the first loading vector. This is coherent with the results observed for BW and local models and the discussion at the end of the previous section.

6.4 Multiphase Approach

The multiphase approach is proposed to overcome the shortcomings of the single model approach with a reduced number of submodels. On the one hand, BW models may not perform adequately when several periods of the batch process are nonlinearly related or independent. This is discussed in Section 6.4.1. On the other hand, as commented before, VW models are poor models facing changing correlation structures. This is further studied in Section 6.4.2. Finally, batch dynamic models can be affected by both nonlinearity or independence and changes in the correlation structure, as it is explained in Section 6.4.3. In the three cases, the model can be improved by independently modeling periods of the batch: the phases.

Let us define a phase as a segment of the batch duration which is well approximated by a single linear model. From this definition it should be stressed that the optimal division into phases of a process depends on the structure of the linear models used. Using BW models, the phases are periods where the variables are linearly related. Using VW models, the phases are periods with a constant correlation structure. Finally, for batch dynamic models, the phases are periods with constant dynamics and where the current and n lags are linearly related.

The multiphase modeling approach provides more flexible models compared to the single model and K-models approaches, but it also has its limitations. First, the use of local linear models for modeling nonlinearities does not take into account part of the multivariate nature of the data – i.e., when approximating nonlinearity with local models, part of the information is lost. Second, the changes in the correlation structure of a batch may be smooth. In that case, smooth transitions between submodels may be more appropriate than a crisp division into phases. Although smooth transitions can be achieved with the currently available modeling techniques, for instance by using the Fuzzy theory or the Mixtures of Probabilistic PCA,[161] it should be noted that these methodologies may complicate the use and interpretation of the models. The use of fuzzy logic for multiphase modeling was originally proposed by Camacho and Picó[26] and has been investigated by Zhao et al.[194] This possibility is not addressed in this document, but it may be a very interesting line of study for the future.

6.4.1 Phases in Batch-wise Data

When data include independent or nonlinearly related groups of variables, a different model should be designed for each of these groups. In Figure 6.13, a pair of examples are shown. Imagine a dataset composed of a hundred of observations of four variables, where the relationship between the first two (see Figure 6.13a) and between the last two (see Figure 6.13b) is perfectly linear. Nonetheless, the first pair of variables is either independent of the last pair (example in Figure 6.13c) or nonlinearly related to it (example in Figure 6.13d).

When the pairs of variables are independent, the analysis can be performed separately for each pair without loss of prediction power and with less number of LVs. For example, a PCA model of the four variables had to include two PCs, whereas the models of the pairs of variables needed only one. Moreover, imagine the second PC is not added to the complete model because its explained variance is low. This can occur with autoscaled data when one of the independent groups of variables has many more variables than the other. In this case, the small group of variables could be poorly modeled. Modeling the groups separately solves the problem.

When dealing with nonlinear relationships, a LVM presents a low prediction power. Figure 6.13d shows a quadratic relationship between variables #2 and #3. The PCA model, which approximates the nonlinear relationship with one PC, produces large residuals and therefore, a high prediction error. If the model is fitted for monitoring, some normal observations can be determined to be abnormal and the other way round, inflating Type I (normal observations detected as faults) and II (nondetected faults) risks. An alternative is to include another PC in the model, but the challenge with the definition of useful control limits would still be present. For instance, in Figure 6.13d, any preprocessed observation which falls between the mean (crossing point of the PCs) and the quadratic curve of circles is abnormal, but the Hotelling's T^2 and SPE commonly used in combination with the PCA model,[119] which are explained in Chapter 7, would catalog it as closer to the average behavior than any of the calibration samples. This happens unless the information obtained from the rest of the relationships among process variables detects the anomaly. Additionally, the model is supposed to be able to predict the values of the first pair of variables from the second pair and the other way round from a linear relationship that does not truly exist.

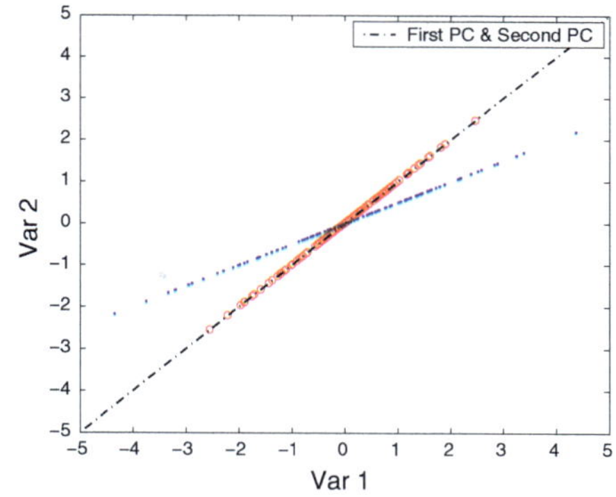

(a) $\{Var1, Var2\}$ space

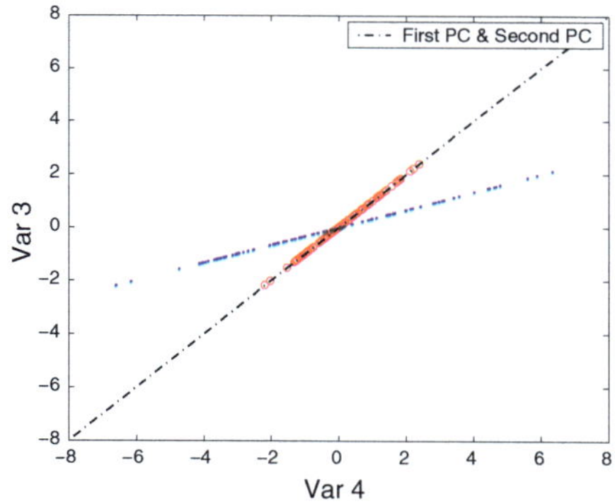

(b) $\{Var3, Var4\}$ space

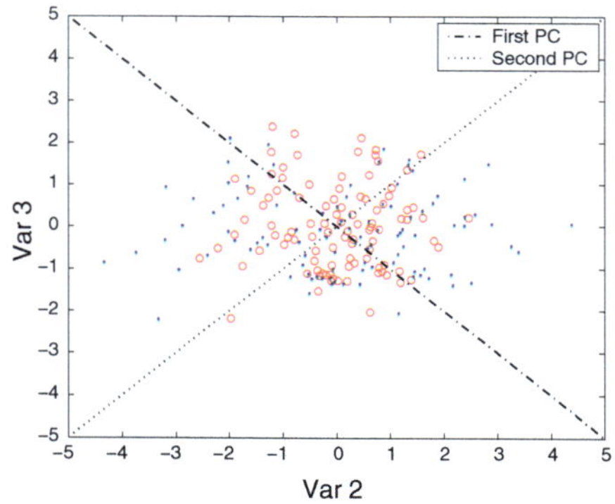

(c) $\{Var2, Var3\}$ space,Independent variables

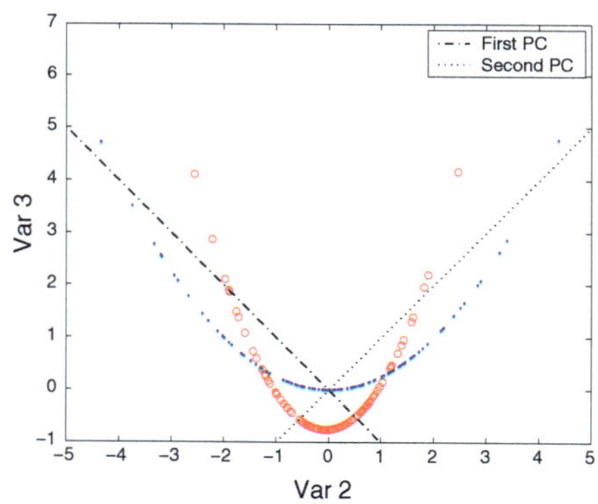

(d) $\{Var2, Var3\}$ space, Nonlinear relationship

Figure 6.13 Relationship between two groups of dependent variables (a and b), and independent or nonlinearly related groups of variables (c and d). Non-preprocessed observations are represented by points. Preprocessed (mean-centered and autoscaled) observations are represented by circles. *Source*: Reproduced from Ref.[26], with permission of Elsevier.

Nonlinear dynamics approximated with a linear combination of variables causes a high prediction and monitoring error in a model of a process. Approximating the nonlinearity with a nonlinear model can be the solution, but this is a difficult task,

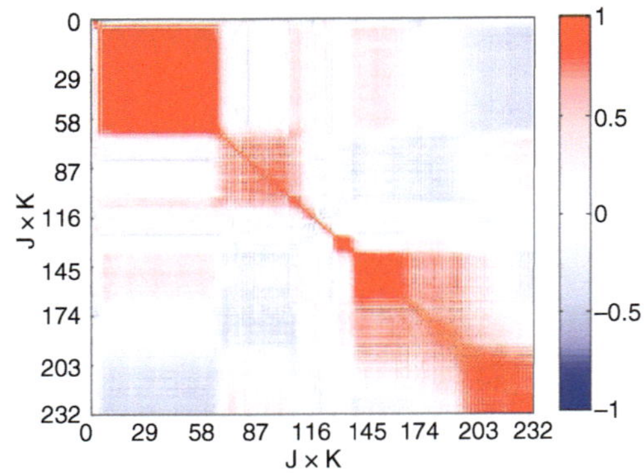

Figure 6.14 Correlation of two variables measured 116 times (batch duration) after BW unfolding and preprocessing (auto-scaling). Nylon 6'6 cultivation process. *Source*: Reproduced from Ref.[28], with permission of John Wiley & Sons.

and a large and rich dataset is needed. In batch processes, a compromising solution is to only model the dynamics which are well approximated by a linear model. Nowadays, as the information of a process is repeated because a large number of variables are collected, a good model of the process can be obtained by modeling the linear or pseudolinear relationships alone. Note that, to improve the prediction power, the multiple models have to be independent. Thus, a multiblock approach[148, 179] should not be applied to model a multiphase process because this approach forces a linear relationship between blocks at the highest level of the model.

Nonlinear relationships and independence between two variables are examples of relationships that are poorly approximated with a linear model. The variables so related show low correlation values. In batch processes, two observations of the vector of process variables commonly show less correlation as the time (grade of completion of the batch) distance between them increases. In Figure 6.14, the correlation map of the BW unfolded data of two variables of the polymerization of nylon 6'6 process is shown. The batch length is 116 and data include the measurements of 50 batches. The result of the BW unfolding is a bidimensional array with 50 observations of 232 variables.

Figure 6.14 shows a diagonal of high correlation. The dark rectangles of the graphic represent the segments of the batch where the variables are highly correlated with their preceding values. The correlation gets negligible as the time distance increases. The formation of rectangles of different sizes in the correlation map means that the autocorrelation and lagged cross-correlation is dependent on the phase of the batch and that the modeling with linear models should be done independently for different phases.

Figure 6.15a illustrates the effect of the division in phases of BW data in the resulting covariance matrix. Part of the relationships are simply not modeled, but notice that this part may show low correlation values (as in the light areas of Figure 6.14). The resulting covariance matrix of the submodel of a phase from sampling time k_i to k_e is presented in Figure 6.15b.

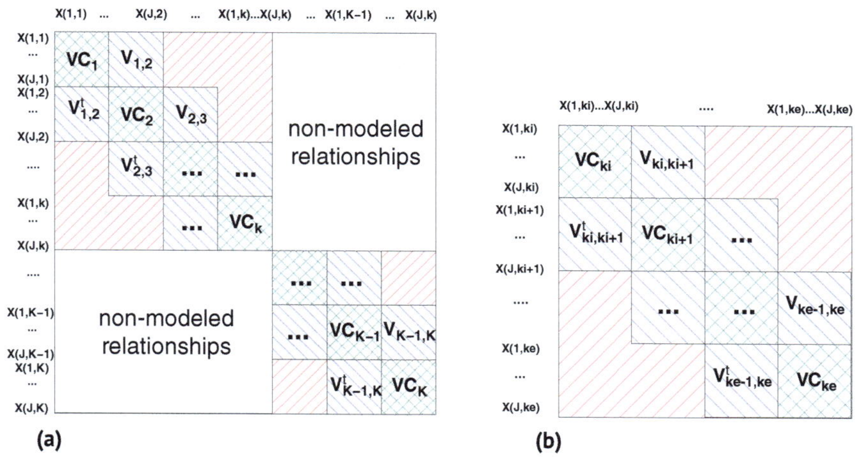

Figure 6.15 (a) Modeled and nonmodeled relationships after a division of a BW model. (b) Covariance matrix of a submodel in the multiphase approach for BW data. *Source*: Reproduced from Ref.[28], with permission of John Wiley & Sons.

Following the notation presented in this chapter, a general multiphase partition from BW unfolded data is represented by:

$$\mathbf{X} = \{\underline{\mathbf{X}}_{k_{il}:k_{el}}^{(k_{el}-k_{il})} : l = 1, \ldots, L\}$$
$$s.t.\ k_{il} \leq k_{el},\ k_{i(l+1)} = k_{el} + 1,\ k_{i1} = 1,\ k_{eL} = K \tag{6.15}$$

It should be highlighted that this approach is a generalization of the BW unfolding. When properly calibrated, multiphase models following Equation 6.15 outperform BW models in terms of prediction error[26] and are especially suited for offline batch process monitoring.[25, 30]

6.4.2 Phases in Variable-wise Data

The VW models have two principal drawbacks: they do not model dynamic information and impose a constant correlation structure.[95] The first drawback can be overcome by adding lags, leading to batch dynamic models. The second can be overcome by using several submodels.[3]

Changing correlation structures from clusters in the score plots. The presence of clusters means that the objects of a cluster represent a different reality than the objects of another cluster. If clusters are not modeled separately, there is a high risk of modeling the intercluster variability instead of the intra-cluster variability. This has negative consequences when the model is used for monitoring or prediction purposes. In Figure 6.16, an example of a dataset with two clusters is shown. The direction of the first PC of the PCA model fitted with the complete dataset is represented by the horizontal arrow. The residuals of this PC are very high and so it is

3 Furthermore, by adding a sufficient number of lags, changing correlation structures are effectively modeled at the expense of overparametrization.

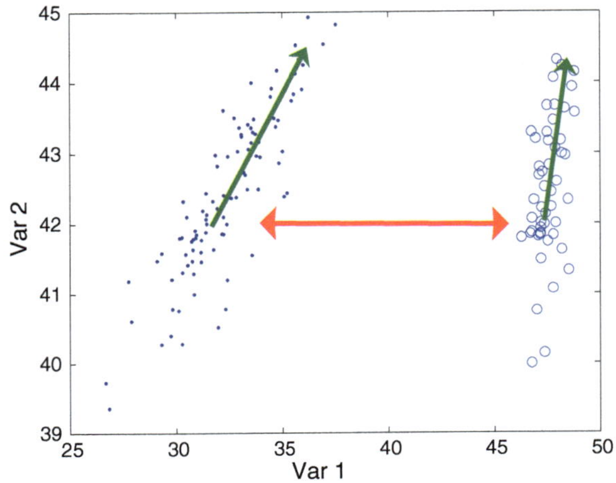

Figure 6.16 Example of a dataset with two clusters. The horizontal arrow represents the first PC fitted for the whole dataset. The other two arrows represent the first PC fitted for each of the two clusters.

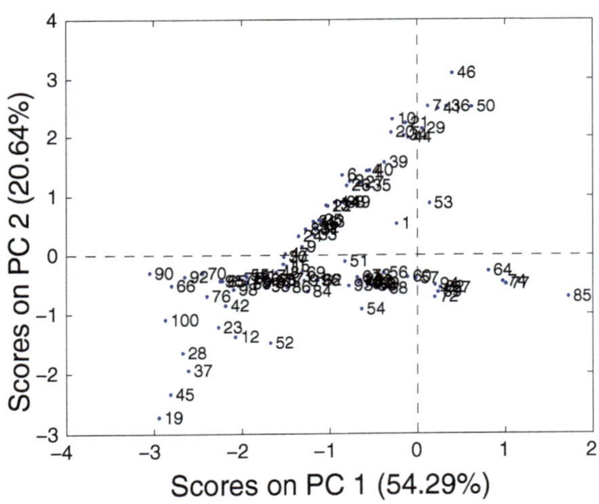

Figure 6.17 Scores of a batch for the 2 first PCs of a VW model from data of the *Saccharomyces cerevisiae* cultivation process. *Source*: Reproduced from Ref.[28], with permission of John Wiley & Sons.

the predictive error. Additionally, this model would have a poor monitoring performance, even if additional PCs were added to it. An improved modeling performance is obtained if the clusters are identified and modeled separately.

In Figure 6.17, the scores of a batch from the *Saccharomyces cerevisiae* cultivation process in a two PCs VW model are shown. It can be seen that the distribution of the scores of the first 50 sampling times (approximately) is very different compared to the last 50 sampling times. These two groups of sampling times should be modeled separately.

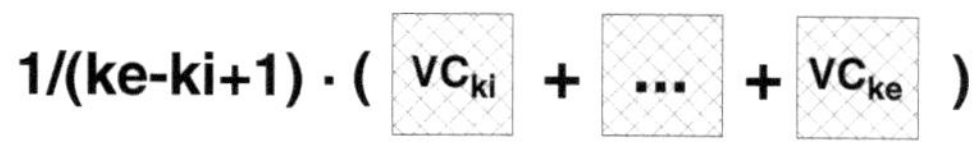

Figure 6.18 Covariance matrix of a submodel in the multiphase approach for VW data. *Source*: Reproduced from Ref.[28], with permission of John Wiley & Sons.

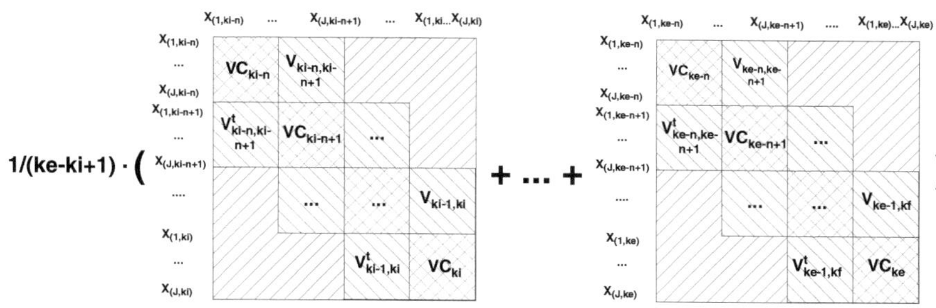

Figure 6.19 Covariance matrix of a submodel in the multiphase approach for batch dynamic data. *Source*: Reproduced from Ref.[28], with permission of John Wiley & Sons.

The resulting covariance matrix of the submodel for a phase from sampling time k_i to k_e is presented in Figure 6.18. The local models of the single sampling times are averaged as in VW unfolding.

A general multiphase partition from VW unfolded data is represented by:

$$\mathbf{X} = \{\underline{\mathbf{X}}^{(0)}_{k_{il}:k_{el}} : l = 1, \ldots, L\}$$
$$s.t.\ k_{il} \le k_{el},\ k_{i(l+1)} = k_{el} + 1,\ k_{i1} = 1,\ k_{eL} = K \tag{6.16}$$

6.4.3 Phases in Batch Dynamic Data

The resulting covariance matrix of the submodel of a phase from sampling time k_i to k_e, for batch dynamic data, is presented in Figure 6.19. The covariance matrices of the UWMW models of the single sampling times – from k_i to k_e – are averaged.

The multiphase models based on batch dynamic data use several submodels for capturing changing correlation structures. The auto-covariances and lagged cross-covariances are modeled by including lags as variables. The size of the window n depends on the amount of past information needed. Each submodel can have a different optimum n value. By properly calibrating this parameter, problems with nonlinear relationships or independence are also avoided.

A general multiphase partition from batch dynamic unfolded data is represented by:

$$\mathbf{X} = \{\underline{\mathbf{X}}^{(n_l)}_{max(k_{il}-n_l,1):k_{el}} : l = 1, \ldots, L\} \tag{6.17}$$

with:

$$n_l = \{k - 1 : k \in \{1, 2, \ldots, k_{el}\}\}$$
$$s.t.\ k_{il} \le k_{el},\ k_{i(l+1)} = k_{el} + 1,\ k_{i1} = 1,\ k_{eL} = K \tag{6.18}$$

Notice that the submodels are not completely disjoint, since the n_l measurement vectors preceding k_{il} are used to fit both submodels $l-1$ and l.

The matrix of Figure 6.18 is a particular case (for $n=0$) of the matrix in Figure 6.19. Nonetheless, a multiphase partition of BW data (see Figure 6.15) cannot be seen as a particular case of the multiphase partition of batch dynamic data. This is a consequence of the definition in Eqs. (6.17) and (6.18). In this definition, the only possibility of obtaining disjoint submodels is for VW unfolding, i.e., $n_l=0$ for $\ell \in \{1, 2, \ldots, L\}$. Therefore, the multiphase partition of BW data in Equation 6.15, which also generates disjoint submodels, is not contemplated in Eqs. (6.17) and (6.18), except for the case of a single-phase BW model.

The definition in Eqs. (6.17) and (6.18) is a general parametrization which includes BW models, VW models, batch dynamic models and UWMW models, being more flexible than any of them. Although in this case, UWMW models have been used to model the phases, an alternative approach is to use EWEW models.

The multiphase models of batch dynamic data are particularly suited for their use online.[27, 29]

6.5 Conclusion

In this chapter, a theoretical discussion regarding the differences among several modeling approaches for batch process data is performed. Approaches are based on bilinear PCA-based modeling and are classified into three main groups: single models, K-models and multiphase approaches.

Single model approaches are those in which the three-way data array is unfolded into a two-way array and then a bilinear PCA-based method is applied. Both BW unfolding and VW unfolding can be seen as special cases of a general unfolding method: the batch dynamic unfolding. In the batch dynamic unfolding, a number of lags are added to the current measurement vector to form the object – row – of the unfolded two-way array. If no LMV is added, the resulting two-way array is the same yielded with VW unfolding. If all possible lags are added, then the unfolded two-way array is the same resulted from the BW unfolding. For the sake of parsimony of the resulting model,[4] the less the number of variables in the unfolded two-way array the better. Also, the inclusion of independent or nonlinearly related variables in the same object is not recommendable for PCA nor for PLS. Therefore, the number of lags in an object should be kept as low as possible, but being high enough to capture the – possibly changing – dynamics. Thus, for a constant correlation structure and when there is little influence of past information in current predictions, the unfolded two-way array should have a low number of lags. When there are changing dynamics or most of the information collected throughout the batch influences the prediction, the appropriate models are more similar to BW models, with many lags included in the unfolded two-way array. In general terms, batch dynamic unfolded

4 Taking into account solely the part of the model applicable to new incoming data. For instance, in PCA the number of parameters associated with the latent structure is understood as the number of elements in the loading matrix and in PLS the number of elements in the regressor matrix.

models are over-parameterized due to this way of modeling time-varying correlation structures.

The K-models approach is based on the calibration of one bilinear PCA-based model per sampling time. This approach is able to capture changing dynamics and avoids problems with nonlinearity and independence between phases of the process. Nonetheless, its principal drawback is the high number of submodels, which makes the calibration and interpretation challenging. This is also the less parsimonious approach and so a larger dataset is needed for proper calibration.

Multiphase models overcome the limitations of single-model approaches with a reduced number of submodels. This is the most flexible and powerful modeling approach of those studied here, as long as the number of submodels and their parameters are properly calibrated taking into account the nature of the data. Therefore, the convenience of this approach relies on the existence of a calibration algorithm for multiphase models.

7

Batch Process Data Analysis and Statistical Monitoring

7.1 Introduction

The final step of the bilinear modeling cycle of batch processes (Chapter 1) resorts to statistical modeling. The four main objectives of batch data analysis[187] are: (i) the analysis of historical batches to gain process understanding and to troubleshoot past abnormal process operating problems, with the ultimate goal of driving the process under normal operating conditions (NOC) again; (ii) the statistical process control (SPC) of incoming batches; (iii) the prediction of final product quality; and (iv) the optimization and/or active control of batch operating conditions to achieve a desired final product quality. This chapter primarily focuses on the first two objectives.

In order to start this modeling step, we assume that the time-varying trajectories of length K for the J variables and/or online analytical sensor responses measured for all the I batches have been equalized, synchronized, arranged into a three-way array $\underline{\mathbf{X}}(I \times J \times K)$, and preprocessed using trajectory C&S. All these steps are usually necessary for extracting the important information in historical data for process understanding and process monitoring (fault detection and diagnosis), as illustrated in the following sections.

7.2 Historical Batch Data Analysis

The analysis of historical data is an important step toward understanding a batch process, including the connections among the following three blocks of data (see Figure 7.1):

- The initial conditions, including starting values for relevant variables – such as initial temperature or pressure, raw material properties, and/or supplier and shift changes of operators. These are arranged into a two-way array $\mathbf{Z}(I \times L)$.
- The aforementioned time-varying process variables and/or online analytical sensor responses were measured throughout the batch evolution and arranged into a three-way array $\underline{\mathbf{X}}(I \times J \times K)$, which are associated with time-varying patterns in batch trajectories.

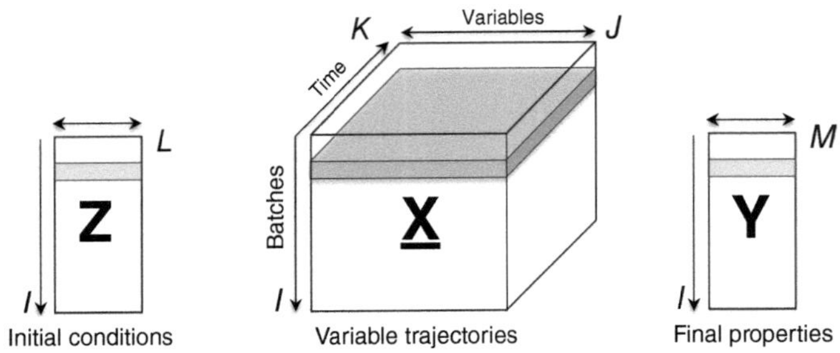

Figure 7.1 Nature of batch-training set data. Each additional prediction-set batch constitutes one additional row in **Z** and one additional layer in **X**. The latter has to begin with only one vector at time = 1 and fills out with one vector per time point during the evolution of the batch. The final quality (**Y**) is usually based on later laboratory analysis of a sample of the product.

- The performance of batches and final product quality arranged into a two-way array $\mathbf{Y}(I \times M)$.

The understanding gained and the models developed at this stage can be used to isolate and diagnose poor operation periods in the past, define an SPC space for the subsequent monitoring phase, and obtain prediction and control models.[187]

There are several approaches for batch data analysis. The simplest one is to extract various meaningful features or landmarks from the trajectories of the time-varying process variables and/or online analytical sensor responses that might be most influential on the process performance. This approach, known as Landmark Feature, extracts this information directly from the three-way array $\underline{\mathbf{X}}(I \times J \times K)$, avoiding the unfolding step. The objective is to summarize the information of the trajectories by extracting various meaningful key characteristics or landmarks from them (e.g., maximum or minimum values, time to the maximum or the minimum value, slope after a particular maximum or minimum values, and duration of certain stages). Once all the H features are extracted from the trajectories, the three-way array $\underline{\mathbf{X}}$ is replaced by a two-way array $\mathbf{F}(I \times H)$ for bilinear modeling. This approach has several drawbacks: (i) a deep understanding of the process is necessary to identify the features that most significantly influence process performance, but this is not always feasible in practice; (ii) a sufficient set of landmark features may not be easily extractable from the variable or batch trajectories (e.g., if trajectories change smoothly along time, distinct features may not be obvious); (iii) important but subtle features may be overlooked; and (iv) features have to be clearly identified in both normal and abnormal batches, and this is not always the case.

More sophisticated but powerful approaches are based on transforming the three-way array $\underline{\mathbf{X}}(I \times J \times K)$ into one or several two-way arrays, using the different unfolding methods (i.e., batch-wise (BW), batch dynamic (BD), or variable-wise (VW) unfolding) and modeling approaches (i.e., single, K-models, adaptive, or multiphase approaches) described in Chapter 6. Among the different options available, a good trade-off between simplicity and modeling capabilities for historical batch data analysis is achieved through BW unfolding. In this way, the three-way array $\underline{\mathbf{X}}(I \times J \times K)$ is transformed into a two-way array $\mathbf{X}(I \times JK)$.

Both in the landmark feature and BW unfolding approaches, the next step is to fit a latent variable model (LVM), such as principal component analysis (PCA), introduced in Chapter 2, to the two-way array $\mathbf{F}(I \times H)$ or $\mathbf{X}(I \times JK)$, respectively. It is also possible to merge the information on initial conditions $\mathbf{Z}(I \times L)$ with either the $\mathbf{F}(I \times H)$ or $\mathbf{X}(I \times JK)$ blocks and calibrate a multi-block PCA.[179] In case of also utilizing the final product quality arranged in a two-way array $\mathbf{Y}(I \times M)$, partial least squares (PLS) models (introduced in Chapter 2) can be used. The combination of BW unfolding and LVM allows us to automatically identify the features of the batch process directly from the historical data, avoiding the limitations of the landmark approach as described before. For the following discussion, we assume that we use BW unfolding.

If a PCA model is fitted, $\mathbf{X}$ is decomposed as:

$$\mathbf{X} = \mathbf{T}_A \mathbf{P}_A^T + \mathbf{E}_A \tag{7.1}$$

where $\mathbf{P}_A$ is the $JK \times A$ loading matrix, with A being the number of latent variables in the model and $A \leq JK$, $\mathbf{T}_A$ is the $I \times A$ score matrix, and $\mathbf{E}_A$ is the $I \times JK$ residual matrix. In this way, each one of the rows of $\mathbf{X}$ associated with the ith batch $\mathbf{x}_i^T(1 \times JK)$ is projected onto the latent space spanned by the columns of $\mathbf{P}$. As a result, the score vector $\boldsymbol{\tau}_i^T(1 \times A)$ (ith row of score matrix $\mathbf{T}_A$)[1] and residual vector $\mathbf{e}_i^T(1 \times JK)$ (ith row of $\mathbf{E}$ residual matrix) are generated.

From the score vector $\boldsymbol{\tau}_i$ and residual vector $\mathbf{e}_i$ associated with the ith row of the two-way array $\mathbf{X}$, two orthogonal and complementary statistics are computed: the Hotelling's T^2 (or D statistic) and squared prediction error (SPE) (or Q statistic).[47] The value of the Hotelling's T_A^2 for the ith observation can be expressed as:

$$T_A^2 = \boldsymbol{\tau}_i^T \boldsymbol{\Theta}^{-1} \boldsymbol{\tau}_i = \sum_{a=1}^{A} \frac{\tau_a^2}{\lambda_a} \tag{7.2}$$

where $\boldsymbol{\Theta}$ $(A \times A)$ is the variance-covariance matrix of the score matrix $\mathbf{T}$. This statistic represents the estimated Mahalanobis distance from the center of the latent subspace to the projection of the batch data onto the A-dimensional subspace. Under the assumption that the scores $\boldsymbol{\tau}_i$ follow a multinormal distribution, since they are linear combinations of random variables, and if a large reference dataset is available, which is typically the case when analyzing historical batches, it holds[163] that T_A^2 (times a constant) follows an F-distribution:

$$T_A^2 \sim \frac{A(I^2-1)}{I(I-A)} F_{A,(I-A)} \tag{7.3}$$

On the other hand, the SPE statistic for the ith batch is given by:

$$\mathrm{SPE} = \mathbf{e}_i^T \mathbf{e}_i = \mathbf{x}_i^T (\mathbf{I} - \mathbf{P}_A \mathbf{P}_A^T) \mathbf{x}_i \tag{7.4}$$

This statistic represents the squared Euclidean distance of the batch from the A-dimensional latent space and provides a measure of how close the batch is to this

1 Note that in the case of PLS, as introduced in Section 2.4.2, the score vector $\boldsymbol{\tau}_i^T(1 \times A)$ (ith row of score matrix $\mathbf{T}_A$) is obtained by projecting the ith batch data vector $\mathbf{x}_i^T(1 \times JK)$ (ith row of matrix $\mathbf{X}$) onto the subspace spanned by the columns of the $JK \times A$ weight matrix $\mathbf{W}_A^*$.

subspace. Assuming that residuals follow a multinormal distribution, Box[16] and Jackson and Mudholkar[83] derived approximate distributions for such quadratic forms. Despite these foundational contributions, traditional residual-based monitoring statistics may exhibit limited performance in high-dimensional or large-scale settings. To address this limitation, Reis et al.[134] introduced the truncated-Q statistic, a refined metric shown to achieve improved detection power, particularly in the context of large-scale systems.

One of the main goals in analyzing historical batch data is to troubleshoot past abnormal process operating problems. For this purpose, two multivariate control charts based on the previously mentioned statistics, T_A^2 and SPE, may be designed by estimating their corresponding control limits. These limits are set at the specific $100(1-\alpha)\%$ percentile of the respective statistical distributions, where α represents the false alarm rate (also termed significance level). An extended practice is to use Shewhart-type control charts for individual observations to monitor both statistics.

Upper control limits for the Shewhart SPE chart can be calculated following several procedures. Jackson and Mudholkar[83] showed that an approximate SPE critical value at false alarm rate α (i.e., $100(1-\alpha)\%$ confidence level) is formulated as follows:

$$\mathrm{UCL(SPE)}_\alpha = \theta_1\left[\frac{z_\alpha\sqrt{2\theta_2 h_0^2}}{\theta_1} + 1 + \frac{\theta_2 h_0(h_0-1)}{\theta_1^2}\right]^{1/h_0} \tag{7.5}$$

where $\theta_k = \sum_{j=A+1}^{\mathrm{rank}(\mathbf{X})}(\lambda_j)^k$, $h_0 = 1 - \frac{2\theta_1\theta_3}{3\theta_2^2}$, λ_j is the jth eigenvalue of PCA residual covariance matrix $\mathbf{E}^T\mathbf{E}/(I-1)$, and z_α is the $100(1-\alpha)\%$ standardized normal percentile. Alternatively, one can use an approximation based on the weighted chi-squared distribution ($g\chi_h^2$) proposed by Box.[16] Nomikos and MacGregor[119] suggested a simple and fast way to estimate the parameters g and h, which is based on matching moments between a $g\chi_h^2$ distribution and the sample distribution of SPE. The mean ($\mu = gh$) and variance ($\sigma^2 = 2g^2h$) of the $g\chi_h^2$ distribution are equated to the sample mean (b) and variance (v) of the SPE sample. Hence, the upper SPE control limit at significance level α is given by:

$$\mathrm{UCL(SPE)}_\alpha = \frac{v}{2b}\chi^2_{(2b^2/v),a} \tag{7.6}$$

where $\chi^2_{(2b^2/v),a}$ is the $100(1-\alpha)\%$ percentile of the corresponding chi-squared distribution.

Regarding Shewhart T_A^2 Hotelling's control chart, the UCL at $100(1-\alpha)\%$ confidence level is given by:

$$\mathrm{UCL}(T_A^2)_\alpha = \frac{A(I^2-1)}{I(I-A)}F_{(A,(I-A)),\alpha} \tag{7.7}$$

where $F_{(A,(I-A)),\alpha}$ is the $100(1-\alpha)\%$ percentile of the corresponding F distribution.

Control charts provide the means to discriminate an unusual batch from NOC representative batches. When the SPE or T^2 values for a particular batch exceed the control limit, it means that this batch is potentially out of control. Hence, a fault diagnosis is required to identify the potential group of variables responsible for such an anomaly. For this purpose, contribution plots[99] are useful tools. A variety of procedures to develop contribution plots can be found in the literature.[53] Here, we describe the most extended approach.

Let us assume that the Shewhart SPE control chart pinpoints the ith batch as abnormal. The contribution of each original jth process variable at the kth sampling time to the high value in the SPE statistic is given by its corresponding squared residual.

$$c_{i,jk}^{\mathrm{SPE}} = e_{i,jk}^2 \tag{7.8}$$

The vector of JK contributions is often depicted as a bar plot where variables and sampling times with high contributions are identified for further investigation.

If the abnormal observation is detected in the Hotelling's T_A^2 control chart, the diagnosis procedure is carried out in two steps: (i) a bar plot of the normalized scores for that batch $(\tau_{i,a}/\lambda_a)^2$ is plotted and the ath score with the highest normalized value is selected; (ii) the contribution of each original jkth variable to this ath normalized score at this new abnormal batch is given by:

$$c_{i,jk}^{T_a^2} = \frac{\tau_{i,a}}{\lambda_a} p_{a,jk} x_{i,jk} \tag{7.9}$$

where $\tau_{i,a}$ is the score value of the ith batch for the ath principal component, $x_{i,jk}$ is the value of the jth process variable at the kth time, and $p_{a,jk}$ is the loading of the jth process variable at the kth time for the ath principal component.[2] As a result, the contributions of all the variables can be displayed to ease fault diagnosis. More precisely, high contributions with the same sign as the score value are subject to investigation, whereas contribution values with the opposite sign should be omitted because these values will only make the score value closer to 0. With the aim of assisting in the fault diagnosis, the exploration of the score scatter plot is recommended, as illustrated in Chapter 2. When there are $V(V \leq A)$ scores with high normalized values, an overall average contribution of the ith batch per variable x_{jk} can be calculated, considering all the selected V scores, as formulated below. For further details, the reader is referred to Refs.[63, 96]

$$\mathrm{CONT}_i(x_{jk}) = \sum_{a=1}^{V} c_{i,jk}^{T_a^2} \tag{7.10}$$

As derived in Eq. (7.10) , the contribution plots provide a list of process variables at particular time periods that contribute numerically to the out-of-control condition due to the deviation from NOCs. Yet, high contributions do not necessarily reveal the actual cause of the fault. Those variables and any other variables highly correlated with them should be investigated. Prior technical process knowledge also plays a crucial role in diagnosing the problem and discovering the root causes of the fault.[63] In fact, leveraging technical process information may assist in identifying an operational action plan to avoid the appearance of the same faults in the future and improve process performance.

---

2 In case of a PLS model, $p_{a,jk}$ is substituted by $w_{a,jk}^*$, the weight of the jkth variable of the batch-wise unfolded two-way array $\mathbf{X}$ in the ath PLS component.

Once the abnormal batches have been detected and diagnosed, new PCA (or PLS) models are calibrated on the set of NOC batches, which constitutes the reference dataset. The benefits of using these new models are twofold: (i) to gain process understanding by exploring the columns of the loading matrix **P** for the different principal components in PCA or the columns of the weight matrix $\mathbf{W}^*$ for the different latent components in PLS; and (ii) to monitoring the performance of future batches using MSPC schemes based on the Hotelling's T_A^2 and SPE statistics, as explained in the following section.

7.3 Batch Multivariate Statistical Process Control

The goal of any SPC scheme is to monitor the performance of a process over time in order to check whether the process behaves as expected when only common-cause variation is present, and to detect any unusual (special or assignable) events that may occur. By finding assignable causes to these anomalies and eliminating their root causes or implementing a performance improvement plan based on their insights, the product quality can be improved. To pursue this goal, control charts are essential tools.[47]

It is important to note that in SPC, the term "control" has a completely different meaning than it has in the control system engineering domain. While in the former it relates to monitoring or surveillance, in the latter it is associated with automatic regulation. On the one hand, in SPC, "control" refers to monitoring the process using data collected from the process to assess the stability of a process over time. The goal is to identify whether the process is in statistical control (i.e., only natural [or common] cause variations are present, and therefore, the process is considered operated under NOC) or out of control (i.e., there are special [or assignable] cause variations that need corrective action). On the other hand, in control system engineering, "control" refers to regulating a process by adjusting manipulable process variables in real time using control algorithms (e.g., PID controllers, cascade controllers, model predictive control) to keep the output at the desired setpoint value. Anyway, both approaches should not be considered alternatives but rather complementary, considering that the monitoring scheme is located at a higher level in the surveillance of the process, being able to monitor the entire system, including the performance of the control algorithms implemented in the process.

The design of statistical monitoring schemes involves two main phases: Phases I and II. Phase I is undertaken using historical data that may contain both good (i.e., in-control) and abnormal (i.e., out-of-control) batches.

7.3.1 Phase I

Phase I starts with the analysis of the PCA (or PLS) model fitted from the historical database. If general assignable causes affecting the whole set of batches are identified, they need to be corrected for, and new historical data need to be retrieved and the model refitted. Once the model is free of general assignable causes, abnormal batches exceeding the control limits of any of the SPE or Hotelling's T_A^2 control charts are detected, diagnosed using contribution plots, and deleted from the database. Then, a new PCA (or PLS) model is built with the remaining batches. This process of model building and model validation is repeated in an iterative way until none of SPE or T_A^2 control charts signal any real out-of-control batch.[3] Then, the PCA (or PLS) model and the SPE or T_A^2 control chart, along with their corresponding control limits for a particular false alarm rate α,[4] built from the good (in-control or NOC) batches that represent the reference dataset, are used for monitoring new batches in the following monitoring stage, Phase II.

A side effect from this debugging procedure is the definition of countermeasures to be implemented in the process operation to avoid the appearance of the same faults in future batches, improving process knowledge and future process performance.

In case of using other unfolding methods (such as BD or VW) or modeling approaches (i.e., K-models, adaptive, or multiphase approaches), the above procedure is applied to the corresponding unfolded two-way array $\mathbf{X}$ discussed in Chapter 6. Note that, for example, in VW unfolding single model, the three-way array $\underline{\mathbf{X}}(I \times J \times K)$ is transformed into a two-way array $\mathbf{X}(IK \times J)$ where each row vector $\mathbf{x}_{ik}^T$ corresponds to the value of the J process trajectories at sampling time k for batch i. If this VW unfolding would be applied for a limited duration of the batch, rather than for its entire duration (as in a multiphase approach), the batch multivariate statistical process control (BMSPC) procedure would be applied to the two-way array corresponding to that period (or phase) $\mathbf{X}(IR \times J)$, where R is the duration of that particular phase of the I batches.

7.3.2 Phase II

Phase II is aimed at assessing and securing process stability by monitoring new (future) batches, once the reference PCA (or PLS) model and the control limits for the SPE or T_A^2 control charts are obtained after Phase I. BMSPC in Phase II can be performed either after batch completion, case in which is called post-batch process monitoring, or during the progress of batch execution, also known as real-time process monitoring.

7.3.2.1 Post-batch Process Monitoring

Post-batch process monitoring can be performed either as end-of-batch (the typical approach) or in a pseudo-online mode, thereby resembling real-time process monitoring. In end-of-batch process monitoring, the analysis of the new batch is performed just as the batch ends. Then, assuming BW unfolding, the new available batch vector $\mathbf{x}_{\text{new}}$, after it has been equalized, synchronized, and preprocessed in a similar way as the reference dataset in the three-way data array $\underline{\mathbf{X}}$, is projected onto the PCA (or PLS) model. As a result, the score vector $\boldsymbol{\tau}_{\text{new}}^T(1 \times A)$ and the residual vector $\mathbf{e}_{\text{new}}^T(1 \times JK)$ are computed, and the values of the Hotelling's T_A^2 and SPE statistics are calculated. This way, the information contained in the original J process

---

3 Note that it is expected that, on average, $100\alpha\%$ of the good (i.e., in-control or NOC) batches lie close above the upper control limits of both SPE or T_A^2 control charts, being α the false alarm rate.

4 Control limits from NOC process data obtained from the theoretical formulas outlined in Eqs. (7.6 and 7.7) can be adjusted using cross-validation techniques to get the required false alarm rate.[31]

variables at K sampling times is summarized in these two indices that are plotted in the corresponding Hotelling's T_A^2 and SPE control charts. These statistics describe the process behavior over the entire duration of the batch. The SPE chart should be checked first. If the values remain below the control limits in both charts, the batch is considered to be in control (i.e., a NOC batch). Upon detecting a value significantly exceeding the upper control limits in any of the control charts, contribution plots, as explained in Section 7.2, are used to identify the specific process variables and time periods that are no longer consistent with normal operating conditions. These contributions provide useful information for assisting in the process diagnosis and root cause analysis to address the potential problems affecting the batch operations and correcting them for subsequent batches. Additionally, the results can be used to separate the product from the batch for storage or further use. This approach is particularly suited for fast processes, such as in semiconductor manufacturing, where a batch takes only a couple of seconds, and there is no time to react during batch evolution, as well as for processes where there is no mechanism for online control or modification of a poor-yield batch after it is detected. To enhance process understanding and detect out-of-control situations, other charts such as univariate time series plots of the scores or scatter score plots can be added to the SPE or Hotelling's T^2 control charts, both in Phase I and II.

In contrast, pseudo-online process monitoring focuses on the statistical evaluation of the process at each sampling time, leading to a more accurate and thorough examination of the process. It is called "pseudo" because it resembles real-time monitoring. In fact, as explained later, both approaches are executed following a similar procedure. The difference between pseudo-online and real-time monitoring is that, in the former, the batch has already finished but is monitored at each sampling time as if it were still running. In the latter, the batch is monitored while it is running and has not finished yet.[65]

In case of using another unfolding method, such as batch dynamic or VW, the above procedure is applied to the new batch after it has been equalized, synchronized, and preprocessed in a similar way as the reference dataset in the three-way data array $\underline{\mathbf{X}}$. This involves using a vector of observations arranged in the same way as the rows of the corresponding unfolded two-way array $\mathbf{X}$ discussed in Chapter 6. For example, in VW unfolding, the new row vector $\mathbf{x}_{\text{new},k}^T(1 \times J)$ to be projected onto the PCA (or PLS) model corresponds to the value of the J process trajectories at time k for the new batch. Considering that, as in end-of-batch process monitoring, all row vectors $\mathbf{x}_{\text{new},k}^T$ for $k = 1$ till K are known, this can be done all at once by projecting all the data registered for the new batch arranged in the two-way array $\mathbf{X}(K \times J)$. In case of using batch dynamic unfolding of order n, which is equivalent to the number of LMVs or lags, the row vector $\mathbf{x}_{\text{new},k}^T(1 \times J(n+1))$ to be projected onto the PCA (or PLS) model corresponds to the value of the J process trajectories from time k till time $n + k$ for the new batch ($k = 1, \dots, K - n$). This can also be done all at once, by projecting the two-way array $\mathbf{X}_{i,n}((K - n) \times J(n + 1))$ formulated in (6.8).

7.3.2.2 Real-time Process Monitoring

Real-time process monitoring focuses on the statistical evaluation of the process as a whole or at several selected sampling times while the new batch is running. In this way, the progress of every new batch can be surveilled, and faults can be detected and potentially diagnosed while the batch is still running. This approach is recommended if one has the ability to act either through online control or some other form of control action made before the end of the batch.[187]

In real-time process monitoring, several factors come into play: (i) a real-time data collection system must be implemented, and (ii) assuming BW unfolding, at a particular time point $k < K$, only the measurements up to that time point are available, and the data for the rest of the batch are missing. Among the several approaches proposed for dealing with this issue, we advocate for the use of missing data imputation methods to fill in, at each time point k, the unknown part of the trajectories from $k + 1$ till K. As commented in Chapter 2, the TSR method[4] is recommended due to its simplicity and good performance. For real-time monitoring of a new batch, several modifications to the scheme used for end-of-batch mode of post-batch process monitoring are required, as explained in the following.

To monitor the progress of a new batch at each time point k, the incomplete batch vector $\mathbf{x}_i^{*T}(1 \times JK)$ has to be equalized in a similar way as the reference dataset and also synchronized. As synchronization has to be performed in real-time, we advocate for the use of the relaxed greedy time warping (RGTW) method, introduced in Chapter 4. Next, the incomplete batch vector will be filled in by using the TSR method, generating a complete batch vector $\mathbf{x}_i^T(1 \times JK)$[5] that, after being preprocessed in a similar way as the reference dataset, is projected onto the PCA (or PLS) model, generating the score vector $\boldsymbol{\tau}_{i,k}^T(1 \times A)$ and residual vector $\mathbf{e}_{i,k}^T(1 \times JK)$. From the score vector $\boldsymbol{\tau}_{i,k}^T$, the Hotelling's $T_{A,k}^2$ statistics shall be calculated and plotted on the corresponding Hotelling's T_A^2 control chart. Unlike in end-of-batch mode or post-batch process monitoring, in real-time process monitoring, the SPE statistic is calculated using the residuals corresponding exclusively to time k for the different process trajectories, rather than the complete residual vector $e_{i,k}^T$.[6]

$$\mathrm{SPE}_k = \sum_{t=(k-1)J+1}^{kJ} e_t^2 \tag{7.11}$$

Note that this procedure is repeated at each time point k ($k = 1, 2, \ldots, K$).

Control limits for a given false alarm rate α, as calculated in Eqs. (7.6 and 7.7), can no longer be applied as they are not appropriate to be used sequentially over the entire duration of the batch. The false alarm rate associated with monitoring the complete K times (full duration of the synchronized batch) will be different from the α value for testing each of the instantaneous times k. Note that as a result of the imputation method, the score vectors $\boldsymbol{\tau}_{i,k}^T$ and residual vectors $\mathbf{e}_{i,k}^T$ obtained at each time point k are not independent. The overall false alarm rate can only be estimated from the joint statistical distribution of the corresponding $T_{A,k}^2$ and SPE_k statistics.

5 Note that the batch vector $\mathbf{x}_{i,k}^T$, obtained at time k, contains the real measurements of the process trajectories up to time point k and the estimated measurements of the process trajectories for the remainder of the batch ($k + 1, \ldots, K$).

6 The reason is that $\mathbf{e}_{i,k}^T$, obtained at time k, does not represent the instantaneous perpendicular distance of a batch from the low-dimensional latent space, and it is affected by the errors associated with the imputation of future unknown part of the process variables trajectories.[119]

An approximate estimation can be obtained by cross-validation using the reference (i.e., in-control or NOC) dataset.[65]

In case of using batch dynamic unfolding of order n, the row vector to be projected at each time k ($k = 1, 2, \dots, K-n$), $\mathbf{x}^T_{\text{new},k}(1 \times J(n+1))$, only has measurements for the J process trajectories at time k, and the rest of the n sampling times from $k+1$ till $k+n$ need to be imputed. Once data are imputed, the resulting vector can be projected onto the PCA (or PLS) model to generate the Hotelling's $T^2_{A,k}$ and SPE_k statistics.

Note that if VW unfolding method is used, then there is no need to impute the future behavior and the row vector $\mathbf{x}^T_{\text{new},k}(1 \times J)$, corresponding to the value of the J process trajectories at time k for the new batch, which is projected into the PCA (or PLS) model to also generate the Hotelling's $T^2_{A,k}$ and SPE_k statistics.

7.4 Practical Issues

From the perspective of the applicability and effectiveness of MSPC approaches based on LVM models, it is important to bear in mind that, as with any inferential approach, two fundamental assumptions must be satisfied for successful implementation: processes must be "comparable" and events of interest must be "observable." The first assumption establishes that the method remains valid as long as the historical reference database is representative of the operation of the process. If anything changes in the process (e.g., the catalyst or the agitation conditions are changed), then it is necessary to collect a new database under these new conditions to apply the method effectively. The second assumption expresses the requirement that the events that one wishes to detect must be "observable" from the measurements that are being collected in the process.

The proposed monitoring schemes, based on PCA and PLS techniques, can also be extended to scenarios where processes can be considered as consisting of a series of blocks. Within these blocks, variables show high correlation, while correlations between variables in different blocks are lower. These blocks may correspond to different physical units of the process, different subprocesses, or distinct sections of the same physical unit, etc. All these sources or blocks of information, together with the two-way arrays **Z** (batch initial conditions) and **Y** (quality variables of the final product), can be integrated into a single BMSPC scheme by using multi-block projection methods.[185] These methods allow the development of monitoring control charts for both blocks of variables and the process as a whole and, in general, reduce the reaction time in the detection of anomalies, while facilitating the diagnosis of the causes of failure. Examples of the use of these multi-block methods in real processes can be found in Refs.[15, 84, 97, 99, 139, 146, 179]

As previously commented, BMSPC can be performed using PCA or PLS models. A common challenge for practitioners is determining what is better to use: PCA on initial conditions **Z**, process variable trajectories **X** and/or landmark features **F**, or PLS, also including the response variables **Y**. Based on the authors' experience, PCA on initial conditions **Z**, process variable trajectories **X**, and/or landmark features **F** is generally a better approach for BMSPC. There are some reasons supporting this statement[94]:

- When developing a model for monitoring purposes (i.e., using only high-quality batches), PLS models usually explain a small percentage of the variability in Y-space (i.e., response variables). This seemingly contradictory result can be explained since, for high-quality batches in general, product variability tends to be low, as product quality is maintained by controlling process conditions.
- PCA identifies the main sources of variability affecting the process operation. Therefore, any deviation from normal operating conditions can be detected in PCA monitoring, even though it is not directly related to product quality disturbances. However, as PLS focuses on the sources of variability in the process related to the product quality, anomalies or process upsets might go undetected in a PLS monitoring system, especially if it does not immediately affect the product quality. The main reason is that the faulty variables might either be excluded or given low weight in the PLS model. Overall, incipient faults are often identifiable by PCA before they affect product quality.
- Sometimes, using PCA to monitor process data is the only viable option, especially when quality data is unavailable at certain stages of the process. In such cases, it would be beneficial to monitor batch process variables to detect anomalies early on, providing an early warning of potential poor batch performance.
- Measurements on critical product quality properties are often infrequent and expensive. In addition, not all product properties are analyzed by the manufacturer. In some situations, the limited number of properties measured on a product may not fully capture its quality and performance across various applications. For example, if only the viscosity of a polymer is measured and kept within specified limits, any variations in end-use applications resulting from differences in chemical structure, such as branching, composition, or end-group concentration, would not be detected. In these cases, the process data may contain more insights about special cause events that could affect the product structure and, consequently, it can be related to the performance in different tertiary applications or processing.

It is crucial to emphasize that the BMSPC outlined in this chapter is designed for data collected while the batch process is being controlled by feedback and feedforward automatic control systems. If an assignable cause arises during a batch, the controllers may compensate for it. While the final product could still meet quality standards, control charts would flag out the batch as abnormal due to the unusual behavior resulting from excessive control actions that could be discovered by using contribution plots. In this way, process operators could take some countermeasures to avoid the occurrence of assignable causes in future batches. Note that this is a very important remark because the assignable cause could be of such magnitude that it would cause a saturation of the control algorithms, which would not be able to compensate for, generating an off-specification product.

Example 7.1: BMSPC

This example illustrates how to design and implement a monitoring control system using a Saccharomyces cerevisiae dataset already synchronized with the RGTW algorithm. As a novelty in this example, we will illustrate the modeling cycle not only with code, which is available in the MATLAB® script "ex_7.mlx," but also using the user interface of the MVBatch toolbox.[7,[65]]

To initiate MVBatch, load the dataset "saccha_NOC" through the Load option. Then, in the screening module (see Figure 7.2), users can visualize raw process variable trajectories for all batches. MVBatch displays calibration data in gray, highlighting the first batch in red. Users can navigate batches via the batch selection panel and access additional variables through the variable selection panel. Based on prior knowledge, variables and batches can be included or excluded using the respective panels.

The next module refers to the alignment where the users can synchronize batch trajectories by interacting with the three-dimensional arrangement panel (see Figure 7.3). As in the code example, the NOC batch data is synchronized with the RGTW. This synchronization is enabled only when a synchronization based on DTW has been previously completed. For this purpose, the geometric average of Kassidas et al. and Ramaker et al.'s weights is applied. Then, the RGTW synchronization is carried out by defining the optimal width of the synchronization window $\zeta = 2$ after using the cross-validation procedure.

Once the batch data are synchronized, the three-way data arrays are preprocessed using the trajectory C&S method (prep=2), and a PCA single-model is fitted using three PCs. For this modeling, the ||`multiphaseFit`|| function is used, which computes the multiphase model parameters for the calibration dataset "cn_alg" and using phase parameters available in the input phases. As a result, the function will return model parameters, such as loadings, scores, means, and standard deviations, as well as the control limits at 95 and 99% confidence interval for D (Hotelling's T^2) and Q (SPE) control charts using both end-of-batch and pseudo-online execution mode.

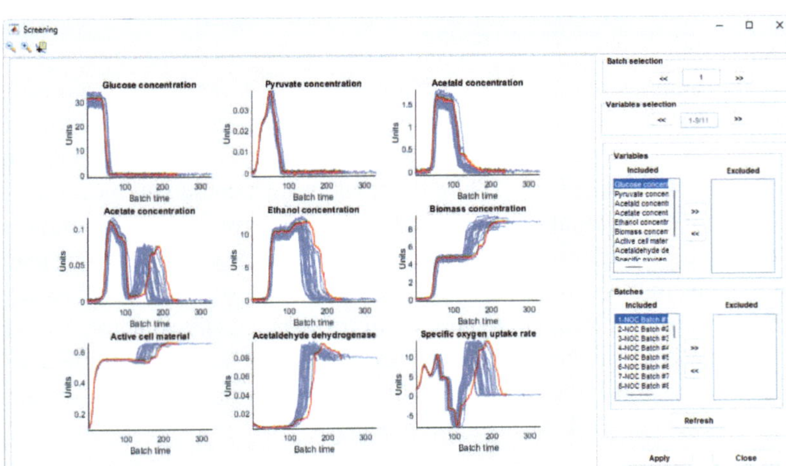

Figure 7.2 Interface for screening batches and variables for subsequent alignment and modeling.

7 https://github.com/jogonmar/MVBatch/releases/tag/1.2.

Figure 7.3 Interface for data synchronization.

Figure 7.4 Interface for data modeling using the multiphase framework.

```
[P,T,mn,stnd,~,onD95,onD99,onQ95,onQ99,offD95,offD99,
    offQ95,offQ99] = multiphaseFit(cn_alg,phases,prep,1);
```

This modeling can be replicated using the Modeling module of the MVBatch user interface, as shown in the "Model" panel in Figure 7.4. The Modeling user interface also provides the option of fitting multiphase models automatically (see the "MPPCA" panel at the center-left in Figure 7.4). This panel would be used to automatically recognize the appropriate model structure for monitoring.

Regarding the design of a monitoring scheme, one recommended step is to re-adjust the theoretical control limits using a leave-one-out cross-validation approach. To execute this procedure, we use the ||`crossvalMVstatistics`|| function, which is equivalent to press the CV monitoring system button in Figure 7.5.

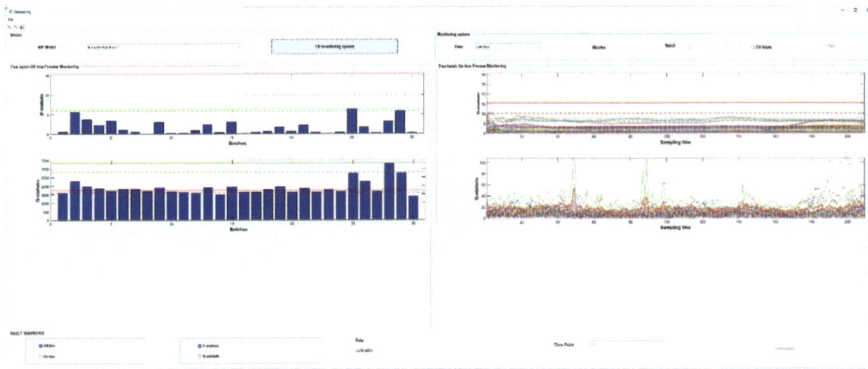

Figure 7.5 Interface for the design of a monitoring scheme and the monitoring of test batches.

```
[cvevolD,cvevolQ,cvD,cvQ,onD95cv,onD99cv,onQ95cv,onQ99cv,
   onD95cv,onD99cv,onalpQ95cv,onalpQ99cv,offD95cv,
   offD99cv,offQ95cv,offQ99cv,offalpD95cv,offalpD99cv,
   offalpQ95cv,offalpQ99cv] = crossvalMVstatistics(cn_alg
   ,phases,prep);
```

Afterward, the "post-batch pseudo-online" control charts for the D statistic, Q statistic, and the warping information can be plotted jointly with their control limits, as illustrated in Figure 7.5 (right). In case that a batch-wise PCA model is fitted, the overall D statistic and Q statistic "post-batch end-of-batch" control charts will also be shown, as illustrated for this example in Figure 7.5 (left). The red control limits are computed using theoretical approximations, and the green control limits are corrected by cross-validation.

To proceed with the monitoring of new batches, we first proceed with the pseudo-online synchronization of each batch using the RGTW algorithm followed by the projection of the resulting synchronized data onto the PCA model as follows:

```
[synTestBatch,warptest] = onSyn(cal{1},Bref,band,
   weights_geom,zeta,rng);
[evolD,evolQ,D,Q,cont_evolD,cont_evolQ] =
   multiphaseProjection(cn_alg,synTestBatch,phases,P,T,mn
   ,stnd,1);
```

Alternatively, users can press the monitor button in the MVBatch user interface to show the D and Q control charts for end-of-batch (left) and pseudo-online (right) monitoring mode, as shown in Figure 7.6.

Figure 7.6 shows that none of the calibration batches exhibit any unusual behavior caused by special-cause process variations.[8] Hence, the monitoring scheme built in Phase I is suitable to assess and secure process stability by monitoring new (future) batches (i.e., Phase II). For this, let us load the test dataset containing 6 faulty batches for exploratory data analysis. To load the dataset "saccha_faults" in the MVBatch

8 Otherwise, abnormal batches should have been diagnosed using contribution plots, and deleted from the database.

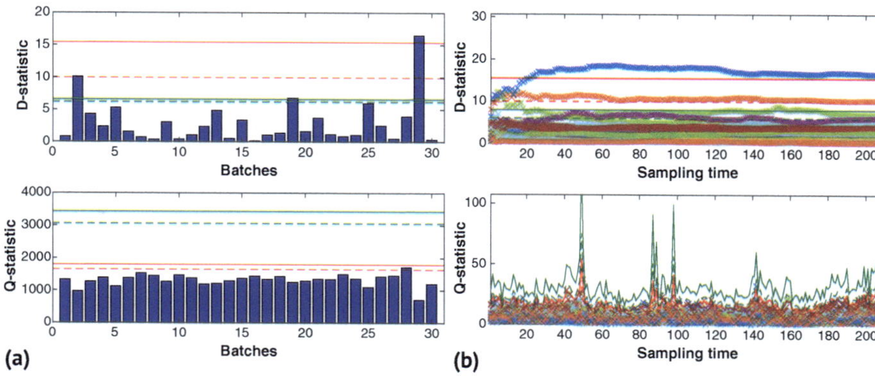

Figure 7.6 Multivariate statistics in Phase I for (a) post-batch end-of-batch and (b) post-batch pseudo-online monitoring.

user interface, users need to select "File" and then click on "Import Data" in the menu bar.

After the test batches have been loaded, they are synchronized, preprocessed, and unfolded based on the monitoring scheme developed in Phase I. Subsequently, the resulting data is projected onto the PCA model, resulting in the multivariate statistics.

In the MVBatch user interface, users can select a batch of the dataset and visualize the multivariate control charts by clicking on the Plot button. To use the corrected control limits in the monitoring of new batches, the CV limits radio button must be ticked on. The "post-batch end-of-batch" control charts are visualized because a batch-wise PCA model has been calibrated (see Figure 7.7a). Furthermore, the "post-batch pseudo-online" control chart for the first batch is also visualized (see Figure 7.7b). Note that only the corrected control limits by cross-validation in the monitoring of new batches are used.

Figure 7.7a shows that the Q statistic values for the first three batches clearly exceed the control limits. Indeed, for the first batch, the "post-batch pseudo-online" (see Figure 7.7b) shows that, from the beginning, the batch is out-of-control based on the Q statistic. Hence, if real-time monitoring had been implemented, this out-of-control signal would have been detected promptly, without the need to wait until the end of the batch.

In cases where out-of-control signals appear in any of the control charts, as is the case in this example, fault diagnosis can provide insights into the potential root causes of these out-of-control signals (see the bottom side of the interface in Figure 7.5). For fault diagnosis, first select the Q statistic and Online radio buttons. To visualize the contributions, the contribution button must be pressed, thereby generating Figure 7.8.

Figure 7.8 shows the contribution plots to the overall Q statistic for the first test batch. This first graph at the top-left side shows the overall contributions to the statistic per variable and throughout the batch run time. The contribution for each variable is enclosed by black dashed lines. The second graph at the center-left side represents the accumulated contribution of each variable through the batch run to the multivariate statistic. Finally, the third graph at the bottom-left side depicts the

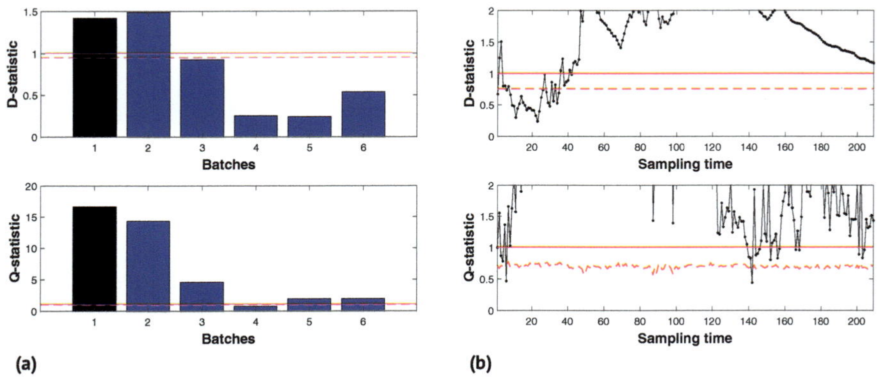

Figure 7.7 Multivariate statistics in Phase II for (a) post-batch end-of-batch and (b) post-batch pseudo-online for first test batch.

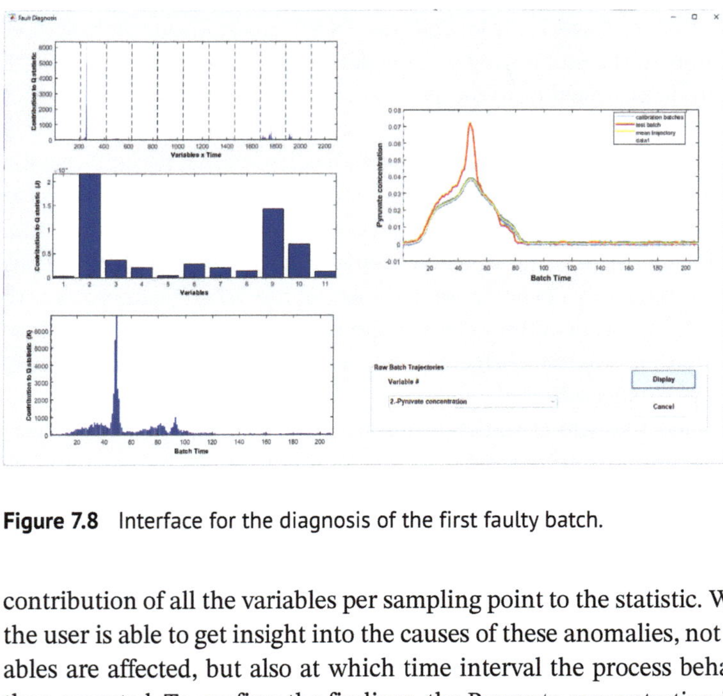

Figure 7.8 Interface for the diagnosis of the first faulty batch.

contribution of all the variables per sampling point to the statistic. With these tools, the user is able to get insight into the causes of these anomalies, not only what variables are affected, but also at which time interval the process behaved differently than expected. To confirm the findings, the Pyruvate concentration is visualized for the calibration batches, the faulty batch, and the average trajectory (see the right side of Figure 7.8).

List of Acronyms

AHKM adaptive hierarchical K-models
ANOVA analysis of variance
BD batch dynamic
BMSPC batch multivariate statistical process control
BW batch-wise
CGF criterium of goodness of fit
CMD correlation matrix distance
COW correlation optimized warping
DTW dynamic time warping
DDTW derivative dynamic time warping
DTW dynamic time warping
EM expectation–maximization
EWEW exponentially weighted evolving window
EWMA exponentially weighted moving average
FDA Food and Drug Administration
GCA grey component analysis
HDDTW hybrid derivative dynamic time warping
IV indicator variable
LMVs lagged measurement vectors
LSD least significant difference
LV latent variable
LVM latent variable model
MCR multivariate curve resolution
MD missing data
MSPC multivariate statistical process control
NIPALS nonlinear iterative partial least squares
NN neural networks
NOC normal operating conditions
NSD normalized squared difference
OLS ordinary least squares

OWU observation wise unfolding
OWU-TBWU observation-wise unfolding-T scores batch-wise unfolding
PARAFAC parallel factor
PAT process analytical technology
PC principal components
PCA principal component analysis
PCR principal component regression
PLS partial least squares
PRESS predicted residual sum of squares
RDDTW robust derivative dynamic time warping
RGTW relaxed greedy time warping
RR ridge regression
SCT stretching, compressing and translating
SIMCA soft independent modelling of class analogy
SPC statistical process control
SPE squared prediction error
SS explained sum of squares
SSR sum of squared residuals
SVD singular value decomposition
TLEC time linear expanding/compression
Trajectory C&S trajectory centering and scaling
TSR trimmed score regression
TSR-MB trimmed score regression model building
UCL upper control limit
UWMW uniformly weighted moving window
Variable C&S variable centering and scaling
VW variable-wise unfolding

Bibliography

1 A. Eleyan and H. Demirel. Face recognition system based on PCA and feedforward neural networks. *Computational Intelligence and Bioinspired Systems. Lecture Notes in Computer Science*, pages 935–942. Berlin, Heidelberg: Springer, 2005.

2 S.W. Andersen and G.C. Runger. Automated feature extraction from profiles with application to a batch fermentation process. *Journal of the Royal Statistical Society: Series C (Applied Statistics)*, 61(2):327–344, 2012.

3 D.T. Andrews and P.D. Wentzell. Applications of maximum likelihood principal component analysis: Incomplete data sets and calibration transfer. *Analytica Chimica Acta*, 350:341–352, 1997.

4 F. Arteaga and A. Ferrer. Dealing with missing data in MSPC: Several methods, different interpretations, some examples. *Journal of Chemometrics*, 16:408–418, 2002.

5 F. Arteaga and A. Ferrer. Framework for regression-based missing data imputation methods in on-line MSPC. *Journal of Chemometrics*, 19:439–447, 2005.

6 F. Arteaga, A. Folch-Fortuny, and A. Ferrer. 2.29 - missing data. In S. Brown, R. Tauler, and B. Walczak, editors, *Comprehensive Chemometrics*, 2nd ed, Oxford, Elsevier, 2020.

7 AspenTech. Analysis of data storage technologies for the management of real-time process manufacturing data, 2015.

8 K.J. Åström. *Introduction to Stochastic Control Theory*. New York: Dover, 2006. Reprint. Originally published by Academic Press 1970.

9 B.R. Bakshi and G. Stephanopoulos. Compression of chemical process data by functional approximation and feature extraction. *AIChE Journal*, 42(2):477–492, 1996.

10 M.S. Bartlett. Tests of significance in factor analysis. *The British Journal of Psychology*, 3:77–85, 1950.

11 H. Bechmann, H. Madsen, N.K. Poulsen, and M.K. Nielsen. Grey box modeling of first flush and incoming wastewater at a wastewater treatment plant. *Environmetrics*, 11:1–12, 2000.

12 T.G. Bloemberg, J. Gerretzen, A. Lunshof, R. Wehrens, and L.M.C. Buydens. Warping methods for spectroscopic and chromatographic signal alignment: A tutorial. *Analytica Chimica Acta*, 781:14–32, 2013.

13 D. Bonne and S.B. Jorgensen. Data-driven modeling of batch processes. In *Proceedings of 7th International Symposium on Advanced Control of Chemical Processes, ADCHEM*, 2004.

14 D. Bonvin. Optimal operation of batch reactors-a personal view. *Journal of Process Control*, 8:355–368, 1998.

15 J. Borràs-Ferrís, C. Duchesne, and A. Ferrer. Multivariate SPC via sequential multiblock-PLS. *Chemometrics and Intelligent Laboratory Systems*, 254:105236, 2024.

16 G.E.P. Box. Some theorems on quadratic forms applied in the study of analysis of variance problems: Effect of inequality of variance in one-way classification. *The Annals of Mathematical Statistics*, 25:290–302, 1954.

17 L. Breiman and J.H. Friedman. Predicting multivariate responses in multiple linear regression. *Journal of the Royal Statistical Society, Series B*, 59(1):3–54, 1997.

18 E.H. Bristol. Swinging door trending: Adaptive trend recording? In *ISA National Conference Proceedings*, pages 749–753. 1990.

19 R. Bro and S. de Jong. A fast non-negativity-constrained least squares algorithm. *Journal of Chemometrics*, 11:393–401, 1997.

20 R. Bro, K. Kjeldahl, A.K. Smilde, and H.A. Kiers. Cross-validation of component models: A critical look at current methods. Analytical and Bioanalytical Chemistry, 390:1241–1251, 2008.

21 R. Bro and A.K. Smilde. Centering and scaling in component analysis. *Journal of Chemometrics*, 17:16–33, 2003.

22 J. Camacho. Missing-data theory in the context of exploratory data analysis. *Chemometrics and Intelligent Laboratory Systems*, 103(1):8–18, 2010.

23 J. Camacho and A. Ferrer. Cross-validation in PCA models with the element-wise k-fold (ekf) algorithm: Theoretical aspects. *Journal of Chemometrics*, 27(7):361–373, 2012.

24 J. Camacho and A. Ferrer. Cross-validation in PCA models with the element-wise k-fold (ekf) algorithm: Practical aspects. *Chemometrics and Intelligent Laboratory Systems*, 131:37–50, 2014.

25 J. Camacho and J. Picó. Monitorización de procesos por lotes mediante PCA multifase. *Revista Iberoamericana De Automatica E Informatica Industrial*, 3:78–91, 2006.

26 J. Camacho and J. Picó. Multi-phase principal component analysis for batch processes modelling. *Chemometrics and Intelligent Laboratory Systems*, 81(2):127–136, 2006.

27 J. Camacho and J. Picó. Online monitoring of batch processes using multi-phase principal component analysis. *Journal of Process Control*, 10(16):1021–1035, 2006.

28 J. Camacho, J. Picó, and A. Ferrer. Bilinear modelling of batch processes. Part I: Theoretical discussion. *Journal of Chemometrics*, 22(5):299–308, 2008.

29 J. Camacho, J. Picó, and A. Ferrer. Bilinear modelling of batch processes. Part II: A comparison of PLS soft-sensors. *Journal of Chemometrics*, 22(10):533–547, 2008.

30 J. Camacho, J. Picó, and A. Ferrer. Multi-phase analysis framework for handling batch process data. *Journal of Chemometrics*, 22:632–643, 2008.

31 J. Camacho, J. Picó, and A. Ferrer. On-line monitoring of batch processes based on PCA: Does the modelling structure matter? *Analytica Chimica Acta*, 642:59–69, 2009.

32 J. Camacho, J. Picó, and A. Ferrer. Data understanding with PCA: Structural and variance information plots. *Chemometrics and Intelligent Laboratory Systems*, 100(1):48–56, 2010.

33 J.D. Carroll and J. Chang. Analysis of individual differences in multidimensional scaling via an n-way generalization of 'eckart-young' decomposition. *Psychometrika*, 45:3–24, 1980.

34 J. Chen and J. Liu. Post analysis on different operating time processes using orthonormal function approximation and multiway principal component analysis. *Journal of Process Control*, 10(5):411–418, 2000.

35 J. Chen and K. Liu. On-line batch process monitoring using dynamic PCA and dynamic PLS models. *Chemical Engineering Science*, 57:63–75, 2002.

36 A. Çinar, S.J. Parulekar, C. Ündey, and G. Birol. *Batch Fermentation: Modeling: Monitoring, and Control*. Chemical Industries. Taylor & Francis, 2003.

37 D. Clifford, G. Stone, I. Montoliu, S. Rezzi, F. Martin, P. Guy, S. Bruce, and S. Kochhar. Alignment using variable penalty dynamic time warping. *Analytical Chemistry*, 81(3):1000–1007, 2009.

38 The National Academy of Sciences Committee on New Sensor Technologies. Expanding the vision of sensor materials, 1995. https://www.nationalacademies.org/read/4782/chapter/1, Accessed 19 December 2025.

39 S. de Jong. SIMPLS: An alternative approach to partial least squares regression. *Chemometrics and Intelligent Laboratory Systems*, 18:251–263, 1992.

40 O.E. de Noord. The influence of data preprocessing on the robustness and parsimony of multivariate calibration models. *Chemometrics and Intelligent Laboratory Systems*, 23(1):65–70, 1994.

41 D. Dong and T.J. McAvoy. Nonlinear principal component analysis-based on principal curves and neural networks. *Computers & Chemical Engineering*, 20(1):65–78, 1996.

42 C. Duchesne, T. Kourti, and J.F. MacGregor. Multivariate SPC for startups and grade transitions. *AIChE Journal*, 48(12):2890–2901, 2002.

43 C. Duchesne and J.F. MacGregor. Multivariate analysis and optimization of process variable trajectories for batch processes. *Chemometrics and Intelligent Laboratory Systems*, 51(51):125–137, 2000.

44 H.T. Eastment and W.J. Krzanowski. Cross-validatory choice of the number of components from a principal component analysis. *Technometrics*, 24(1):73–77, 1982.

45 P.H.C. Eilers. Parametric time warping. *Analytical Chemistry*, 76(2):404–411, 2004.

46 L. Eriksson, E. Johansson, N. Kettaneh-Wold, J. Trygg, C. Wikström, and S. Wold. *Multi- and Megavariate Data Analysis*. MKS Umetrics AB, 2006.

47 A. Ferrer. Multivariate statistical process control based on principal component analysis (MSPC-PCA): Some reflections and a case study in an autobody assembly process. *Quality Engineering*, 19(4):311–325, 2007.

48 S. Feyodeazevedo, B. Dahm, and F. Oliveira. Hybrid modelling of biochemical processes: A comparison with the conventional approach. *Computers & Chemical Engineering*, 21:S751–S756, 1997.

49 A. Folch-Fortuny, F. Arteaga, and A. Ferrer. PCA model building with missing data: New proposals and a comparative study. *Chemometrics and Intelligent Laboratory Systems*, 146:77–88, 2015.

50 I.E. Frank and J.H. Friedman. A statistical view of some chemometrics regression tools. *Technometrics*, 35:109–148, 1993.

51 P.M. Frank. Fault diagnosis in dynamic systems using analytical and knowledge based redundancy: A survey and some new results. *Automatica*, 26:459–474, 1990.

52 M. Fransson and S. Folestad. Real-time alignment of batch process data using COW for on-line process monitoring. *Chemometrics and Intelligent Laboratory Systems*, 84:56–61, 2006.

53 M. Fuentes-García, G. Maciá-Fernández, and J. Camacho. Evaluation of diagnosis methods in PCA-based multivariate statistical process control. *Chemometrics and Intelligent Laboratory Systems*, 172:194–210, 2018.

54 K.R. Gabriel. The biplot graphic display of matrices with application to principal component analysis. *Biometrika*, 58:453–467, 1971.

55 S. García-Muñoz, T. Kourti, and J.F. MacGregor. Troubleshooting of an industrial batch process using multivariate methods. *Industrial and Engineering Chemistry Research*, 42:3592–3601, 2003.

56 S. García-Muñoz, T. Kourti, and J.F. MacGregor. Model predictive monitoring for batch processes. *Industrial and Engineering Chemistry Research*, 43(18):5929–5941, 2004.

57 S. García-Muñoz, M. Polizzi, A. Prpich, C. Strain, A. Lalonde, and V. Negron. Experiences in batch trajectory alignment for pharmaceutical process improvement through multivariate latent variable modelling. *Journal of Process Control*, 21(10):1370–1377, 2011.

58 P. Geladi and B.R. Kowalski. Partial least-squares regression: A tutorial. *Analytica Chimica Acta*, 185:1–17, 1986.

59 D. Giancarlo and C. Tommasi. Cross-validation methods in principal component analysis: A comparison. *Statistical Methods and Applications*, 11:71–82, 2002.

60 G. Gins, P. Van den Kerkhof, and J.F.M. Van Impe. Hybrid derivative dynamic time warping for online industrial batch-end quality estimation. *Industrial & Engineering Chemistry Research*, 51(17):6071–6084, 2012.

61 G. Gins, J. Espinosa, I. Y. Smets, W. Van Brempt, and J.F.M. Van Impe. Data alignment via dynamic time warping as a prerequisite for batch-end quality prediction. In *Proceedings of the 6th Industrial Conference on Data Mining Conference on Advances in Data Mining: Applications in Medicine, Web Mining, Marketing, Image and Signal Mining*, ICDM'06, pages 506–510, Berlin, Heidelberg: Springer-Verlag, 2006.

62 K. Gollmer and C. Posten. Supervision of bioprocesses using a dynamic time warping algorithm. *Control Engineering Practice*, 4(9):1287–1295, 1996.

63 J.M. González-Martínez. Advances on bilinear modeling of biochemical batch processes. PhD thesis, Universidad Politecnica de Valencia, 2015.

64 J.M. González-Martínez, J. Camacho, and A. Ferrer. Bilinear modeling of batch processes. Part III: Parameter stability. *Journal of Chemometrics*, 28:10–27, 2014.

65 J.M. González-Martínez, J. Camacho, and A. Ferrer. MVbatch: A matlab toolbox for batch process modeling and monitoring. *Chemometrics and Intelligent Laboratory Systems*, 183(15):122–133, 2018.

66 J.M. González-Martínez, O.E. de Noord, and A. Ferrer. Multisynchro: A novel approach for batch synchronization in scenarios of multiple asynchronisms. *Journal of Chemometrics*, 28:462–475, 2014.

67 J.M. González-Martínez, A. Ferrer, and J.A. Westerhuis. Real-time synchronization of batch trajectories for on-line multivariate statistical process control using dynamic time warping. *Chemometrics and Intelligent Laboratory Systems*, 105:195–206, 2011.

68 J.M. González-Martínez, A. Ferrer, and J.A. Westerhuis. Relaxed-greedy time warping: A new tool for real-time synchronization of batch trajectories for batch MSPC. In *Proceedings of the 5th International Chemometrics Research Meeting*. Nijmegen (The Netherlands), 2011.

69 J.M. González-Martínez, A. Folch-Fortuny, F. Llaneras, M. Tortajada, J. Picó, and A. Ferrer. Metabolic flux understanding of pichia pastoris grown on heterogeneous culture media. *Chemometrics and Intelligent Laboratory Systems*, 134:89–99, 2014.

70 J.M. González-Martínez, R. Vitale, O.E. de Noord, and A. Ferrer. Effect of synchronization on bilinear batch process modeling. *Industrial & Engineering Chemistry Research*, 53:4339–4351, 2014.

71 J.M. González-Martínez, J.A. Westerhuis, and A. Ferrer. Using warping information for batch process monitoring and fault classification. *Chemometrics and Intelligent Laboratory Systems*, 127:210–217, 2013.

72 S.P. Gurden, J.A. Westerhuis, S. Bijlsma, and A.K. Smilde. Modelling of spectroscopy batch process data using grey models to incorporate external information. *Journal of Chemometrics*, 15:101–121, 2001.

73 S.P. Gurden, J.A. Westerhuis, R. Bro, and A.K. Smilde. A comparison of multiway regression and scaling methods. *Chemometrics and Intelligent Laboratory Systems*, 59 (1–2):121–136, 2001.

74 R.A. Harshman. Foundations of the PARAFAC procedure: Models and conditions for an 'explanatory' multimodal factor analysis. *UCLA Working Papers in Phonetics*, 16:1–84, 1970.

75 H. He and X. Yu. A comparison of PCA/ICA for data preprocessing in remote sensing imagery classification. In D. Li and H. Ma, editors, *MIPPR 2005: Image Analysis Techniques*, pages 60–65. Proceedings of the SPIE, volume 6044 2005.

76 I.S. Helland. Partial least squares regression and statistical models. *Scandinavian Journal of Statistics*, 17:97–114, 1990.

77 I.S. Helland. On the structure of partial least squares. *Communication in Statistics - Simulation and Computation*, 17(2):581–607, 1988.

78 M. Herdin, N. Czink, H. Ozcelik, and E. Bonek. Correlation matrix distance, a meaningful measure for evaluation of non-stationary mimo channels. In *Vehicular Technology Conference, 2005. VTC 2005-Spring. 2005 IEEE 61st*, volume 1 pages 136–140, 2005.

79 L.E. Holloway and B.H. Krogh. On-line trajectory encoding for discrete-observation process monitoring. In *Proceedings of IFAC Symposium On-line Fault Detection and Supervision in the Chemical Process Industries*, 1992.

80 A. Hoskuldsson. PLS regression methods. *Journal of Chemometrics*, 2:211–228, 1988.

81 A. Hyvärinen and E. Oja. Independent component analysis: Algorithms and applications. *Neural Networks*, 13(4–5):411–430, 2000.

82 J.E. Jackson. *A User's Guide to Principal Components*. England, Wiley-Interscience, 2003.

83 J.E. Jackson and G.S. Mudholkar. Control procedures for residuals associated with principal component analysis. *Technometrics*, 21:341, 1979.

84 C. Kiparissides, J.F. MacGregor, C. Jaeckle, and M. Koutoudi. Process monitoring and diagnosis by multiblock PLS methods. *AIChE Journal*, 40:826–838, 1994.

85 I.T. Jolliffe. *Principal Component Analysis*. Springer, 2002.

86 N. Kaitsha and C.F. Moore. Extraction of event times in batch profiles for time synchronization and quality predictions. *Industrial & Engineering Chemistry Research*, 40(1):252–260, 2001.

87 A. Kassidas. Fault detection and diagnosis in dynamic multivariable chemical processes using speech recognition methods. *Open Access Dissertations and Theses Paper* 3386, 1997.

88 A. Kassidas, J.F. MacGregor, and P.A. Taylor. Synchronization of batch trajectories using dynamic time warping. *AIChE Journal*, 44:864–875, 1998.

89 K.J. Kauffman, P. Prakash, and J.S. Edwards. Advances in flux balance analysis. *Current Opinion in Biotechnology*, 14(5):491–496, 2003.

90 E.J. Keogh and M.J. Pazzani. Derivative dynamic time warping. In *In First SIAM International Conference on Data Mining (SDM2001)*, 2001.

91 K.B. Konstantinov and T. Yoshida. A method for on-line reasoning about the time-profiles of process variables. In *Proceedings of IFAC Symposium On-line Fault Detection and Supervision in the Chemical Process Industries*, 1992.

92 K.A. Kosanovich, K.S. Dahl, and M.J. Piovoso. Improved process understanding using multiway principal component analysis. *Industrial & Engineering Chemistry Research*, 35:138–146, 1996.

93 T. Kourti. Abnormal situation detection, three-way data and projection methods; robust data archiving and modeling for industrial applications. *Annual Reviews in Control*, 27:131–139, 2003.

94 T. Kourti. Multivariate dynamic data modeling for analysis and statistical process control of batch processes, start-ups and grade transitions. *Journal of Chemometrics*, 17:93–109, 2003.

95 T. Kourti. Process analysis and abnormal situation detection: From theory to practice. *IEEE Control Systems Magazine*, 22(5):10–25, 2003.

96 T. Kourti. Application of latent variable methods to process control and multivariate statistical process control in industry. *International Journal of Adaptive Control and Signal Processing*, 19:213–246, 2005.

97 T. Kourti, J. Lee, and J.F. MacGregor. Experiences with industrial applications of projection methods for multivariate statistical process control. *Computers & Chemical Engineering*, 20:S745–S750, 1996.

98 T. Kourti and J.F. MacGregor. Process analysis, monitoring and diagnosis, using multivariate projection methods. *Chemometrics and Intelligent Laboratory Systems*, 28:3–21, 1995.

99 T. Kourti and J.F. MacGregor. Multivariate SPC methods for process and product monitoring. *Journal of Quality Technology*, 28(4):409–428, 1996.

100 W. Ku, R.H. Storer, and C. Georgakis. Disturbance detection and isolation by dynamic principal component analysis. *Chemometrics and Intelligent Laboratory Systems*, 30(1):179–196, 1995.

101 S. Lakshminarayanan, R. Gudi, and S. Shah. Monitoring batch processes using multivariate statistical tools: Extensions and practical issues. In *Proceedings of lFAC Worm Congress*, pages 241–246, 1996.

102 B. Lennox, G.A. Montague, H.G. Hiden, G. Kornfeld, and P.R. Goulding. Process monitoring of an industrial fed-batch fermentation. *Biotechnology and Bioengineering*, 74:125, 2001.

103 Y-J. Liu, S. Andre, L. Saint Cristau, S. Lagresle, Z. Hannas, E. Calvosa, O. Devos, and L. Duponchel. Multivariate statistical process control (MSPC) using raman spectroscopy for in-line culture cell monitoring considering time-varying batches synchronized with correlation optimized warping (COW). *Analytica Chimica Acta*, 952:9–17, 2017.

104 D.J. Louwerse and A.K. Smilde. Multivariate statistical process control of batch processes based on three-way models. *Chemical Engineering Science*, 55:1225–1235, 1999.

105 N. Lu, F. Gao, and F. Wang. Sub-PCA modeling and on-line monitoring strategy for batch processes. *AIChE Journal*, 50(1):255–259, 2004.

106 N. Lu, F. Gao, Y. Yang, and F. Wang. PCA-based modeling and on-line monitoring strategy for uneven-length batch processes. *Industrial and Engineering Chemical Research*, 43:3343–3352, 2004.

107 J.F. MacGregor and T. Kourti. Statistical process control of multivariate processes. *Control Engineering Practice*, 3(3):403–414, 1995.

108 O. Marjanovic, B. Lennox, D. Sandoz, K.Smith, and M. Crofts. Real-time monitoring of an industrial batch process. *Computers & Chemical Engineering*, 30 (10–12):1476, 2006.

109 H. Martens and T. Næs. *Multivariate Calibration*. John Wiley & Sons, 1992.

110 M.D. Mesarovic, S.N. Sreenath, and J.D. Keene. Search for organising principles: Understanding in systems biology. *Systems Biology, IEE*, 1(1):19–27, 2004.

111 C.E. Miller. *Chemometrics in Process Analytical Chemistry*, pages 226–328. Blackwell Publishing Ltd, 2007.

112 M. Misra, S. Joe Qin, S. Kumar, and D. Seemann. On-line data compression and error analysis using wavelet technology. *AIChE Journal*, 46(1):119–132, 2000.

113 Z.K. Nagy and R.D. Braatz. Nonlinear model predictive control for batch processes. In W.S. Levine, editor, *The Control Handbook*, Volume Control System Advance Methods, Chapter 15, pages 1–23. CRC Press, 2011.

114 P.R.C. Nelson, P.A. Taylor, and J.F. MacGregor. Missing data methods in PCA and PLS: Score calculations with incomplete observations. *Chemometrics and Intelligent Laboratory Systems*, 35:45–65, 1996.

115 D. Neogi and C. Schlags. Multivariate statistical analysis of an emulsion batch process. *Industrial and Engineering Chemistry Research*, 37:3971–3979, 1998.

116 C.W. Ng and M.A. Hussain. Hybrid neural network – Prior knowledge model in temperature control of a semi-batch polymerization process. *Chemical Engineering and Processing: Process Intensification*, 43(4):559–570, 2004.

117 N. Nielsen, J. Carstensen, and J. Smedsgaard. Aligning of single and multiple wavelength chromatographic profiles for chemometrics data analysis using correlation optimised warping. *Journal of Chromatography*, 805:17–35, 1998.

118 P. Nomikos and J.F. MacGregor. Monitoring batch processes using multiway principal component analysis. *AIChE Journal*, 40:1361–1375, 1994.

119 P. Nomikos and J.F. MacGregor. Multivariate SPC charts for monitoring batch processes. *Technometrics*, 37(1):41–59, 1995.

120 H. Nyquist. Certain topics in telegraph transmission theory. *Transactions of the American Institute of Electrical Engineers*, 47(2):617–644, 1928.

121 U.S. Department of Health, Human Services, Food and Drug Administration, Center for Drug Evaluation, Research (CDER), Center for Veterinary Medicine (CVM), Office of Regulatory Affairs (ORA), and Pharmaceutical CGMPs. Guidance for industry, PAT: A framework for innovative pharmaceutical development, manufacturing, and quality assurance, 2004. https://www.fda.gov/media/71012/download, Accessed 19 December 2025.

122 M. Onel, C.A. Kieslich, Y.A. Guzman, C.A. Floudas, and E.N. Pistikopoulos. Reprint of: Big data approach to batch process monitoring: Simultaneous fault detection and diagnosis using nonlinear support vector machine-based feature selection. *Computers & Chemical Engineering*, 116:503–520, 2018.

123 OSIsoft. The compression algorithm – kb00699, 2015.

124 K. Pearson. On lines and planes of closest fit to points in space. *Philosophical Magazine*, 2:559–572, 1901.

125 J. Pettersson and P. Gutman. Automatic tuning of the window size in the box car back slope data compression algorithm. *Journal of Process Control*, 14(4):431–439, 2004.

126 V. Pravdova, B. Walczak, and D. Massart. A comparison of two algorithms for warping of analytical signals. *Analytica Chimica Acta*, 456:77–92, 2002.

127 H. Ramaker, E.N.M. van Sprang, S.P. Gurden, J.A. Westerhuis, and A.K. Smilde. Improved monitoring of batch processes by incorporating external information. *Journal of Process Control*, 12:569–576, 2002.

128 H. Ramaker, E.N.M. van Sprang, J.A. Westerhuis, and A.K. Smilde. Dynamic time warping of spectroscopic batch data. *Analytica Chimica Acta*, 498:133–153, 2003.

129 H. Ramaker, E.N.M. van Sprang, J.A. Westerhuis, and A.K. Smilde. Fault detection properties of global, local and time evolving models for batch process monitoring. *Journal of Process Control*, 15(7):799–805, 2005.

130 J.O. Ramsay and B.W. Silverman. *Functional Data Analysis*. New York: Springer-Verlag, 1997.

131 S. Rännar, J.F. MacGregor, and S. Wold. Adaptive batch monitoring using hierarchical PCA. *Chemometrics and Intelligent Laboratory Systems*, 41:73–81, 1998.

132 T.J. Rato and M.S. Reis. An integrated multiresolution framework for quality prediction and process monitoring in batch processes. *Journal of Manufacturing Systems*, 57:198–216, 2020.

133 T.J. Rato, J. Blue, J. Pination, and M.S. Reis. Translation-invariant multiscale energy-based PCA for monitoring batch processes in semiconductor manufacturing. *IEEE Transactions on Automation Science and Engineering*, 14:894–904, 2017.

134 M.S. Reis, R. Rendall, T.J. Rato, C. Martins, and P. Delgado. Improving the sensitivity of statistical process monitoring of manifolds embedded in high-dimensional spaces: The truncated-q statistic. *Chemometrics and Intelligent Laboratory Systems*, 215:104369, 2021.

135 R. Rendall, L.H. Chiang, and M.S. Reis. Data-driven methods for batch data analysis – A critical overview and mapping on the complexity scale. *Computers & Chemical Engineering*, 124:1–13, 2019.

136 R. Rocha de Oliveira and A. de Juan. Synchronization-free multivariate statistical process control for online monitoring of batch process evolution. *Frontiers in Analytical Science*, 1, 2022.

137 S.G. Rothwell, E.B. Martin, and A.J. Morris. Comparison of methods for handling unequal length batches. In *Proceedings of IFAC DYCOPS5, Corfu, Greece*, pages 66–71, 1998.

138 C. Runge. Über empirische funktionen und die interpolation zwischen äquidistanten ordinaten. *Zeitschrift für Mathematik und Physik*, 46(224–243):20, 1901.

139 N. Kettaneh, S. Wold, and K. Tjessem. Hierarchical multiblock PLS and PC models for easier model interpretation and as an alternative to variable selection. *Journal of Chemometrics*, 10:463–482, 1996.

140 E. Saccenti and J. Camacho. On the use of the observation-wise k-fold operation in PCA cross-validation. *Journal of Chemometrics*, 29(8):467–478, 2015.

141 H. Sakoe and S. Chiba. Dynamic programming algorithm optimization for spoken word recognition. *IEEE Transactions on Acoustics, Speech, and Signal Processing*, 26(1):43–49, 1978.

142 J. Sansana, M. N. Joswiak, I. Castillo, Z. Wang, R. Rendall, L. H. Chiang, and M. S. Reis. Recent trends on hybrid modeling for industry 4.0. *Computers & Chemical Engineering*, 151:107365, 2021.

143 F. Sartori, P. Facco, F. Zuecco, F. Bezzo, and M. Barolo. Optimal indicator-variable approach for trajectory synchronization in uneven-length multiphase batch processes. *Industrial & Engineering Chemistry Research*, 62(44):18511–18525, 2023.

144 C.E. Shannon. Communication in the presence of noise. *Proceedings of the IRE*, 37(1):10–21, 1949.

145 A. Simoglou, P. Georgieva, E.B. Martin, A.J. Morris, and S. Feyo de Azevedo. On-line monitoring of a sugar crystallization process. *Computers & Chemical Engineering*, 29:1411–1422, 2005.

146 S. Valle, S.J. Qin, and M.J. Piovoso. On unifying multiblock analysis with application to decentralized process monitoring. *Journal of Chemometrics*, 15:715–742, 2001.

147 J. Sjöblom, O. Svensson, M. Josefson, H. Kullberg, and S. Wold. An evaluation of orthogonal signal correction applied to calibration transfer of near infrared spectra. *Chemometrics and Intelligent Laboratory Systems*, 44:229–244, 1998.

148 A.K. Smilde, J.A. Westerhuis, and S. de Jong. A framework for sequential multiblock component methods. *Journal of Chemometrics*, 17(6):323–337, 2003.

149 A.K. Smilde. Comments on three-way analysis used for batch process data. *Journal of Chemometrics*, 15:19–27, 2001.

150 A.K. Smilde, R. Bro, and P. Geladi. *Multi-way Analysis, Application in the Chemical Sciences*. England: John Wiley & Sons, 2003.

151 Sohlberg. Grey box modelling for model predictive control of a heating process. *Journal of Process Control*, 13:225–238, 2003.

152 M. Spooner and M. Kulahci. Monitoring batch processes with dynamic time warping and k-nearest neighbours. *Chemometrics and Intelligent Laboratory Systems*, 183:102–112, 2018.

153 R. Srinivasan and M.S. Qian. Off-line temporal signal comparison using singular points augmented time warping. *Industrial & Engineering Chemistry Research*, 44(13):4697–4716, 2005.

154 R. Srinivasan and M.S. Qian. Online fault diagnosis and state identification during process transitions using dynamic locus analysis. *Chemical Engineering Science*, 61(18):6109–6132, 2006.

155 R. Srinivasan and M.S. Qian. Online temporal signal comparison using singular points augmented time warping. *Industrial & Engineering Chemistry Research*, 46(13):4531–4548, 2007.

156 R. Steuer and B.H. Junker. Computational Models of Metabolism: Stability and Regulation in Metabolic Networks. New York: Wiley, 2008.

157 Y. Takane, H.A.L. Kiers, and J. de Leeuw. Component analysis with different sets of constraints on different dimensions. *Psychometrica*, 60:259–280, 1995.

158 Y. Takane and T. Shibayama. Principal component analysis with external information on both subjects and variables. *Psychometrica*, 56:97–120, 1991.

159 R. Tauler. Multivariate curve resolution applied to second order data. *Chemometrics and Intelligent Laboratory System*, 30:133–146, 1995.

160 N.F. Thornhill, M.A.A.S Choudhury, and S.L. Shah. The impact of compression on data-driven process analyses. *Journal of Process Control*, 14(4):389–398, 2004.

161 M.E. Tipping and C.M. Bishop. Mixtures of probabilistic principal component analyzers. *Neural Computation*, 11:443–482, 1999.

162 G. Tomasi and F. van den Berg. Correlation optimized warping and dynamic time warping as preprocessing methods for chromatographic data. *Journal of Chemometrics*, 18:231–241, 2004.

163 N.D. Tracy, J.C. Young, and R.L. Mason. Multivariate control charts for individual observations. *Journal of Quality Technology*, 24(2):88–95, 1992.

164 L.R. Tucker. The extension of factor analysis to three-dimensional matrices. In N. Frederiksen and H. Gulliksen, editors, *Contributions to Mathematical Psychology*, pages 110–162. New York: Holt, Rinehart and Winston, 1964.

165 UMETRICS. SIMCA 13.0.3. Available at www.umetrics.com, accessed in 2014.

166 C. Ündey and A. Çinar. Statistical monitoring of multistage, multiphase batch processes. *IEEE Control Systems Magazine*, 22(5):40–52, 2002.

167 C. Ündey, S. Ertunç, and A. Çinar. Online batch/fed-batch process performance monitoring, quality, prediction, and variable-contribution analysis for diagnosis. *Industrial & Engineering Chemistry Research*, 42:4645–4658, 2003.

168 C. Ündey, B.A. Williams, and A. Çinar. Monitoring of batch pharmaceutical fermentations: Data synchronization, landmark alignment, and real-time monitoring. In *15th IFAC World Congress*. volume 35 pages 271–276, 2002.

169 E.N.M. van Sprang, H. Ramaker, J.A. Westerhuis, S.P. Gurden, and A.K. Smilde. Critical evaluation of approaches for on-line batch process monitoring. *Chemical Engineering Science*, 57:3979–3991, 2002.

170 E.N.M. van Sprang, H.J. Ramaker, J.A. Westerhuis, A.K. Smilde, and D. Wienke. Statistical batch process monitoring using gray models. *AIChE Journal*, 51(3):931–945, 2005.

171 W.F. Velicer. Determining the number of components from the matrix of partial correlations. *Psychometrika*, 41:321–327, 1976.

172 A. Wachs and D.R. Lewin. Improved PCA methods for process disturbance and failure identification. *AIChE Journal*, 45(8):1688–1700, 1999.

173 B. Walczak and D.L. Massart. Dealing with missing data: Part I. *Chemometrics and Intelligent Laboratory Systems*, 58:15–27, 2001.

174 J. Wan, O. Marjanovic, and B. Lennox. Uneven batch data alignment with application to the control of batch end-product quality. *ISA Transactions*, 53(2):584–590, 2014.

175 M.J. Watson, A. Liakopoulos, D. Brzakovic, and C. Georgakis. A practical assessment of process data compression techniques. *Industrial & Engineering Chemistry Research*, 37(1):267–274, 1998.

176 P.D. Wentzell and L.V. Montoto. Comparison of principal components regression and partial least squares regression through generic simulation of complex mixtures. *Chemometrics and Intelligent Laboratory Systems*, 65:257–279, 2003.

177 F. Westad, L. Gidskehaug, B. Swarbrick, and G.R. Flaten. Assumption free modeling and monitoring of batch processes. *Chemometrics and Intelligent Laboratory Systems*, 149:66–72, 2015.

178 J.A. Westerhuis, E.P.P.A Derks, H.C.J Hoefsloot, and A.K. Smilde. Grey component analysis. *Journal of Chemometrics*, 21:474–485, 2007.

179 J.A. Westerhuis, T. Kourti, and J.F. MacGregor. Analysis of multiblock and hierarchical PCA and PLS models. *Journal of Chemometrics*, 12(5):301–321, 1998.

180 J.A. Westerhuis, T. Kourti, and J.F. MacGregor. Comparing alternative approaches for multivariate statistical analysis of batch process data. *Journal of Chemometrics*, 13 (3–4):397–413, 1999.

181 B.M. Wise, N.B. Gallagher, R. Bro, J.M. Shaver, W. Windig, and R.S. Koch. *PLSToolbox 3.5 for Use with Matlab*. Eigenvector Research Inc., 2005.

182 B.M. Wise, N.B. Gallagher, S.W. Butler, D.D. White, and G.G. Barna. A comparison of principal component analysis, multiway principal component analysis, trilinear decomposition and parallel factor analysis for fault detection in a semiconductor etch process. *Journal of Chemometrics*, 13 (3–4):379–396, 1999.

183 H. Wold and E. Lyttkens. Nonlinear iterative partial least squares (NIPALS) estimation procedures. In Proceedings of Bulletin of the International Statistical Institute, *37th session, London*, pages 1–15. 1969.

184 S. Wold. Cross-validatory estimation of the number of components in factor and principal components. *Technometrics*, 20(4):397–405, 1978.

185 S. Wold, P. Geladi, K. Esbensen, and J. Ohman. Multi-way principal components and PLS analysis. *Journal of Chemometrics*, 1:41–56, 1987.

186 S. Wold, N. Kettaneh, H. Friden, and A. Holmberg. Modelling and diagnostics of batch processes and analogous kinetic experiments. *Chemometrics and Intelligent Laboratory Systems*, 44:331–340, 1998.

187 S. Wold, N. Kettaneh-Wold, J.F. MacGregor, and K.G. Dunn. 4.12 - Batch process modeling and MSPC. In S. Brown, R. Tauler, and B. Walczak, editors, *Comprehensive Chemometrics*, 2nd ed, pages 305–322, Oxford: Elsevier, 2009.

188 S. Wold, L. M. Sjöström, and L. Eriksson. PLS-regression: A basic tool of chemometrics. *Chemometrics and Intelligent Laboratory Systems*, 58:109–130, 2001.

189 M. Zarzo and A. Ferrer. Batch process diagnosis: PLS with variable selection versus block-wise PCR. *Chemometrics and Intelligent Laboratory Systems*, 73:15–27, 2004.

190 P. Zhang. Model selection via multifold cross validation. *The Annals of Statistics*, 21:299–313, 1993.

191 Y. Zhang, M. Dudzic, and V. Vaculik. Integrated monitoring solution to start-up and run-time operations for continuous casting. *Annual Reviews in Control*, 27(2):141–149, 2003.

192 Y. Zhang and T. F. Edgar. A robust dynamic time warping algorithm for batch trajectory synchronization. In *Proceedings of American Control Conference*, pages 2864–2869, 2008.

193 Y. Zhang, B. Lu, and T.F. Edgar. Batch trajectory synchronization with robust derivative dynamic time warping. *Industrial & Engineering Chemistry Research*, 52(35):12319–12328, 2013.

194 C. Zhao, F. Wang, N. Lu, and M. Jia. Stage-based soft-transition multiple PCA modeling and on-line monitoring strategy for batch processes. *Journal of Process Control*, 17(9):728–741, 2007.

195 J. Zhao, W. Li, H. Qu, G. Tian, and Y. Wei. Real-time monitoring and fault detection of pulsed-spray fluid-bed granulation using near-infrared spectroscopy and multivariate process trajectories. *Particuology*, 53:112–123, 2020.

Index

Note: Page numbers in *italics* and **bold** refers to figures and tables respectively.